NATO
Looks South

New Challenges and New Strategies in the Mediterranean

Ian O. Lesser

Prepared for the United States Air Force

Project AIR FORCE • RAND

Preface

The security environment facing the United States and NATO in Europe is continuing to change in fundamental ways almost a decade after the fall of the Soviet Union. One significant change has been the steady growth of security challenges emanating from Europe's southern periphery—around the Mediterranean and beyond. The United States remains the dominant security actor in this region, and NATO strategy is beginning to look more closely at the management of problems outside the center of Europe. European, Middle Eastern, and Eurasian security are becoming interdependent as a result of political, economic, and military trends. The evolution of the strategic environment along these lines has important implications for defense planning, including the future of U.S. and allied air power. It also suggests a growing role for key allies in NATO's south—Spain, Italy, and Turkey—and the growing significance of U.S. and United States Air Force Europe (USAFE) relationships with these countries. The recent Kosovo experience underscores these realities.

This report explores the strategic environment on NATO's southern periphery, with special attention to transregional risks, Turkey's Alliance role, the Mediterranean dimension of NATO adaptation, and what these issues might mean for U.S. strategy and the USAF. This research was undertaken as part of a 1998 project on "Change and Adaptation in NATO: Implications for the USAF," conducted within the Strategy and Doctrine Program of RAND's Project AIR FORCE.

Other reports in this series address: emerging security issues in Europe's east; the strategic environment in the Caucasus and Central Asia (Richard Sokolsky and Tanya Charlick-Paley, *NATO and Caspian Security: A Mission Too Far?* MR-1074-AF, 1999); prospective allied defense contributions and the outlook for European security and defense cooperation; the air power contributions of NATO's new and potential members; and the defense planning and air power implications of a changing NATO.

Project AIR FORCE

Project AIR FORCE, a division of RAND, is the Air Force federally funded research and development center (FFRDC) for studies and analysis. It provides the Air Force with independent analyses of policy alternatives affecting the development, employment, combat readiness, and support of current and future aerospace forces. Research is performed in four programs: Aerospace Force Development; Manpower, Personnel, and Training; Resource Management; and Strategy and Doctrine.

Contents

Summary

THE SOUTHERN PERIPHERY AND EUROPEAN SECURITY

Recent conflicts from Kosovo to Iraq have focused attention on risks emanating from NATO's southern periphery. The April 1999 Washington summit deepened this interest by identifying the Mediterranean as an area of special security concern and reaffirming the commitment to the existing Mediterranean Initiative. The thrust of Alliance strategy, however, is defined in *functional* rather than geographic terms, with an emphasis on new missions from countering proliferation to crisis management, all of which are most likely to be performed in the south.

Risks in the new strategic environment are transregional. European, Middle Eastern, and Eurasian security are increasingly interwoven, and Europe will be more exposed to the consequences of Western policies outside of Europe. To the extent that the Alliance directs its efforts to the defense of common interests and power projection, additional attention will be paid to Southern Region members, the Mediterranean states involved in partnership and dialogue with NATO, and the wider region where developments can affect transatlantic security. An evolution in this direction will also serve U.S. strategic interests, encouraging greater European involvement in defense on the periphery, bolstering the relevance of the U.S. military presence in and around Europe, and contributing to U.S. freedom of action in extra-European crises.

Key security relationships require redefinition. Algeria, Bosnia, and successive crises in the Gulf have played seminal roles in shaping security perceptions around the Mediterranean and transatlantic cooperation. But key U.S. security relationships, both bilateral and through NATO, have not adjusted to reflect post–Cold War realities. These relationships require redefinition to provide a predictable basis for cooperation in addressing post–Cold War problems. NATO's new

Strategic Concept is helpful in this regard, but is not enough. The United States needs to explore ways of jointly redefining key bilateral relationships in the Southern Region through more frequent high-level interaction with leaderships.

The need for redefinition is most acute in Turkey. Internal uncertainties and multiple security risks (the term "threats" still has relevance for Ankara) make Turkey the new front-line state within the Alliance. However, there is no transatlantic consensus on policy toward Turkey, and longer-term uncertainty in Ankara's relations with the European Union (EU) places greater pressure on the bilateral relationship with Washington. Turkey has emerged as a far more important, but also much more difficult security partner. In the absence of a concerted effort to reengage Ankara in European security affairs and to reassure Turkey about the solidity of the NATO security commitment, the United States and the Alliance risk losing a key asset in shaping the new strategic environment. A new agenda for security relations with Ankara will need to focus on proliferation risks, counterterrorism, and energy security—common interests across the Southern Region. It will also need to address Turkey's special concerns about pressure from a resurgent Russia.

Failure to address the risk of a Greek-Turkish conflict jeopardizes Alliance adaptation and European security. Full implementation of risk-reduction measures, along the lines brokered in 1998 by NATO's Secretary General, is imperative. Strategic dialogue to manage longer-term risks, including disputes in the Aegean and Cyprus, should have a broad agenda and might embrace arms control, Balkan and Black Sea reconstruction, and regional crisis management. As a hedge, however, it is essential that the Alliance—or at least key members—develop plans in advance to monitor and contain a possible clash in the eastern Mediterranean.

Expanded NATO involvement in the Mediterranean—Europe's "near abroad"—is a logical step toward a broader transatlantic security partnership. There is more support for this within the Alliance than for more ambitious models of strategy toward common security interests in the Gulf, the Caspian, and elsewhere. Germany is emerging as a significant actor in the Mediterranean region, and can be a part of this evolution. The return of France as a full NATO partner would be a transforming development in strategy toward the south, and should be a priority objective of U.S. policy.

Greater attention to the south in NATO strategy should imply a commensurate shift of Alliance resources. Most, and the most likely, NATO contingencies are in the south, but the vast bulk of Alliance resources remain north of the Alps. Costs associated with the integration of new members in the east will impose competing demands, and an expeditionary strategy may offset requirements for permanently based assets in the south (there may even be benefits to keeping a relatively large proportion of forces in the rear but available for use on the periphery). At a minimum, however, missions in the south, especially counterproliferation and air defense, will require improvements to the undercapitalized and outdated NATO infrastructure across the Southern Region.

More capable allies—for limited, nearby contingencies. NATO's southern allies are in the process of restructuring and modernizing their militaries to create readily deployable forces. Progress on European Security and Defense Identity (ESDI) and NATO's Defense Capabilities Initiative should give further impetus to this trend, although with the exception of Turkey and Greece, low levels of defense spending will place limits on future capabilities. The scale of Turkish modernization plans suggests that it will emerge as a very capable, regional military power over the next decade. The southern allies are already capable of making significant contributions to amphibious and other operations in their own subregions (e.g., in North Africa and around the Adriatic). At the same time, and as the Bosnia and Kosovo operations show, the political will exists to use these forces in regional contingencies.

IMPLICATIONS FOR MILITARY STRATEGY AND AIR POWER

Distance, diversity of risks, and Alliance geography give aerospace power a special role on Europe's southern periphery. The AFSOUTH[1] area of regard now stretches from Mauritania and the

---

[1] Allied Forces Southern Europe (AFSOUTH) is one of two regional commands of NATO's Allied Command Europe (ACE).

Canaries to the Caucasus. The extent of this security space, and the need for NATO to move toward a greater power projection orientation, suggests that the role of air power in European security has changed significantly. European-based air power will likely be called on to a greater extent for interventions outside Europe—including in the Middle East and Eurasia. NATO in the new strategic environment is likely to place more, not less, emphasis on air power, and the bulk of future demands across a range of missions—from humanitarian assistance to countering weapons of mass destruction (WMD) and halting conventional aggression, to counterterrorism and crisis management—will emanate from the south.

Spain, Italy and Turkey will be key to supporting expeditionary operations in the south. This analysis does not suggest the need for significant rebasing of United States Air Force Europe (USAFE) air assets.[2] Rather, an expeditionary approach to power projection in NATO's south suggests the importance of reinforcing access arrangements around the Mediterranean. Italy, and above all, Turkey, will be key centers for the projection of air power in the new environment. Italy's proximity to the Balkans and North Africa, and a generally favorable political acceptance climate, gives it a special role in facilitating the projection of tactical air power, as well as in supporting airlift and strategic air operations further afield.[3] The Kosovo experience reinforces this point.

Turkey will be critical. Its importance in the power projection equation will only be enhanced by future concerns about the Caspian, counterproliferation demands, and possible disruptions in traditional approaches to defense in the Gulf. Moreover, key contingencies for the Alliance could involve the defense of Turkey itself. There will be a need for a framework to allow the long-term rotational presence of tactical air power at Incirlik, or elsewhere, beyond Operation Northern Watch. Above all, the USAF access and overflight relationship must be more

---

[2] See the companion report by David Ochmanek, *NATO's Future: Implications for U.S. Military Capabilities and Posture*, RAND, MR-1162-AF, 2000.

[3] Similar conclusions were reached in the context of a 1991 RAND study. See Ian Lesser and Kevin Lewis, *Airpower and Security in NATO's Southern Region: Alternative Concepts for a USAF Facility at Crotone*, N-3264-AF. The increase in U.S. air traffic through Sigonella provides additional testimony to the importance of Italy as a logistical line to the Gulf.

predictable. Improved military-to-military cooperation can play a role. But translating Turkey's geostrategic advantages, including new alignments with Israel and Jordan, into operational benefits can only be accomplished through high-level bilateral agreement on regional defense strategies.

Preserving traditional military-to-military ties will require new engagement efforts. The United States has close historical ties to southern European and Turkish air forces. The preference for American equipment, training, and security assistance has encouraged close military-to-military relations. The end of U.S. security assistance to NATO Southern Region countries is evidence of more mature relationships, but it also removes a primary basis for cooperation. New defense-industrial initiatives can play this role where arms transfers are noncontroversial. With key countries such as Greece and Turkey, this is not often the case.

Generational change in Southern Region air forces also raises questions about the future of military-to-military cooperation. An increasingly European orientation across southern Europe has encouraged closer defense-industrial and training ties with European partners. In Turkey, the impetus for diversification comes from concerns about the unpredictability of U.S. arms transfers. New engagement efforts through USAFE can help to offset these changes in attitude about bilateral cooperation over the longer term. In broad terms, however, the trend toward diversification may now be a permanently operating factor.

A portfolio approach to presence and access is desirable. Expeditionary approaches to power projection and crisis management on NATO's southern periphery place a premium on flexibility in access. As the political scene in NATO's south (and among countries outside the Alliance but within the NATO orbit) changes, there will be new opportunities for establishing presence and access relationships. Beyond providing additional operational flexibility and extending air power's reach to new areas of concern such as the Black Sea, a portfolio approach can increase the predictability of cooperation by reducing the perception that an ally is being "singularized" (this has been a concern in Italy and Turkey). Candidates for augmenting the portfolio include

Hungary, Romania, and Azerbaijan. Changing attitudes in Greece may also make Crete attractive for North African contingencies. Existing British bases on Cyprus might be useful in relation to the Levant and the Gulf. A portfolio approach to access arrangements is a useful hedge against uncertainties about coalition behavior in future crises, not least, the potential unavailability of transit through the Suez Canal and the resulting increase in airlift requirements.

Increasing the NATO content of air power activities will facilitate cooperation. Where appropriate, existing bilateral air power activities in the south should be given a NATO flavor. NATO content can improve political acceptance and may help accustom southern allies to more expansive Alliance missions. Outside NATO, and especially with NATO's Mediterranean Initiative partners, some bilateral exercises and other activities might also be conducted "in the spirit of" the Mediterranean Initiative. Morocco, Egypt, Israel, and Jordan are key candidates.

Greek-Turkish risk reduction is an imperative, especially in the air. Greece and Turkey possess highly capable air forces, and this capability is set to grow. At the same time, the confrontation over the Aegean and Cyprus increasingly takes place in the air. Initiatives aimed at risk reduction and confidence building in the air can therefore make a disproportionate contribution to stability in the eastern Mediterranean. The political consequences of a Greek-Turkish clash could, for example, include the open-ended denial of access to Turkish bases. Given the nature of the stakes, the United States and NATO should be prepared to contribute air power assets to demilitarization and confidence-building activities, including the monitoring of no-fly zones that might be agreed as part of future Cyprus or Aegean arrangements.

Toward a Southern Strategy for NATO

NATO has taken some steps to integrate Mediterranean security concerns and initiatives in its broader strategy. Given the security demands emanating from the region, a focused strategy toward the south is required, embracing:

- **Core objectives.** The Alliance continues to have important Article V responsibilities in the south, particularly on Turkey's borders. Deterring and defending against these risks to Alliance territory are core objectives of NATO's strategy. A third and increasingly prominent core objective will be to defend common interests on Europe's periphery.
- **Environment shaping.** To help promote NATO's core objectives, NATO strategy needs to address emerging security problems around the Mediterranean in a proactive manner. Key tasks in this regard include the prevention and management of regional crises, including dangerous flashpoints in the Balkans. Similarly, the Alliance needs to contain new security risks of a transregional character such as WMD and missile proliferation, spillovers of terrorism and political violence, and threats to energy security. NATO's Mediterranean Initiative can play an important role in environment shaping by promoting security dialogue and engaging nonmember states in North Africa and the Middle East in defense cooperation, training, and crisis management activities.
- **Hedging against uncertainty.** The Mediterranean is a crisis-prone region experiencing rapid change. NATO strategy must anticipate the need to mitigate the effects of unforeseen crises, including consequences that may be felt on NATO territory. Dealing with disastrous refugee flows and civil emergencies, and preparing for humanitarian interventions, will be key tasks.

AREAS FOR FUTURE RESEARCH AND ANALYSIS

This study indicates where additional research and analysis will be useful. Developments in the Balkans, the Aegean, and the Middle East may offer opportunities to extend U.S. and NATO capabilities and develop more predictable security relationships on Europe's periphery. Key issues for future analysis include: the lessons of Kosovo for USAF basing and access, especially with regard to Italy; the role of U.S. forces operating from Turkey beyond Operation Northern Watch;

and approaches to developing and implementing specific USAF-sponsored risk-reduction programs for Greece and Turkey.

Acknowledgments

The research presented here benefited greatly from discussions with official and unofficial observers in the United States and Europe, including NATO Headquarters and AFSOUTH. The author wishes to thank all those who contributed their views and expertise. He is particularly grateful to the study sponsors, General John Jumper and General Don Peterson, and to project action officers Major Jerry Gandy (USAFE/XPXS) and Captain Don Shaffer (AF/XOOX) for their interest and assistance. Thanks are also due to Andrew Pierre at the U.S. Institute of Peace, and to RAND colleagues James Thomson, David Gompert, Zalmay Khalilzad, Greg Treverton, Stephen Larrabee, Robert Levine, C. Richard Neu, Marten Van Heuven, Robert Hunter, David Ochmanek, Richard Sokolsky, Tanya Charlick-Paley, Peter Ryan, Thomas Szayna, Michele Zanini, Jeanne Heller, and Rosalie Heacock for their comments and assistance on this report and the study as a whole. Needless to say, any errors or omissions are the responsibility of the author.

Chapter 1
Introduction

NATO's southern periphery—the Mediterranean basin together with the Black Sea and its hinterlands—is attracting growing attention in transatlantic security debates, for tangible reasons. The most likely, and some of the most dangerous, security risks in post–Cold War Europe are to be found in the south rather than in the center of the continent. Crises in Algeria, the Balkans, the Levant, and potentially in Cyprus and the Aegean are emblematic of these concerns. At the same time, changes on the political and economic scene have transformed NATO's "Southern Region," and have made southern Europe and Turkey more assertive actors in security affairs and more significant defense partners for the United States.[1]

The adaptation of the Alliance in terms of missions as well as membership reinforces the importance of the south. To the extent that NATO continues to evolve in the direction of the defense of common interests and crisis management in addition to the defense of members' territory, the Mediterranean region—Europe's near abroad—is a natural area for expanded cooperation. The notion of a more "global" alliance remains highly controversial. But the idea of doing more in and around the Mediterranean is now part of the consensus within NATO, and has been strongly reinforced by the Kosovo crisis. For the Alliance, and above all for the United States, a more active stance toward the south is also part of the growing emphasis on power projection and the employment of European-based forces for extra-European contingencies.

Command reforms, NATO's new Strategic Concept, and new avenues for NATO's Mediterranean Initiative can contribute to the ef-

---

[1] NATO's Southern Region traditionally comprises Portugal, Spain, Italy, Greece, and Turkey. Hungary's accession to the Alliance will formally add a sixth southern region member to Allied Forces Southern Europe (AFSOUTH).

fectiveness of the Alliance in addressing southern risks.[2] Despite signs of détente in the eastern Mediterranean, one of NATO's most serious potential flashpoints—relations between Greece and Turkey—remains a challenge for Alliance adaptation. An open conflict between Athens and Ankara could prove disastrous to the future of the Alliance and would severely complicate U.S. strategy in the region. More broadly, Europe and the United States have yet to come to grips with the critical question of how to keep Turkey—a leading "consumer" of security in the new environment but also potentially an important defense partner for the West—positively engaged in European security.

The future of relations with Russia will also be closely tied to developments in the south, from the Balkans and eastern Mediterranean to the Caspian and the Gulf. A more assertive, nationalistic Russia may find it easier to challenge Western interests on the periphery rather than seeking to change the post–Cold War order in central and eastern Europe. Russian policy toward southeastern Europe and the Gulf, including destabilizing arms and technology transfers, and friction over Kosovo could be harbingers of more difficult relations. Moreover, Moscow's own security concerns are increasingly focused on instability to Russia's south—in the Caucasus and Central Asia. The future security of Turkey, as well as the viability of conventional arms control regimes, could be strongly affected by these trends.

COMMON INTERESTS IN THE SOUTH

Alliance security interests in the south can be described in three dimensions. First, NATO has a strong stake in developments emanating from the Mediterranean-Black Sea region as part of the new European security environment. Key issues in this dimension range from soft security concerns such as migration, environmental risks, and fears about civilizational frictions to more tangible worries about new energy dependencies across the Mediterranean. Further along the spectrum of risks, there is a growing but still surprisingly muted (at least in Europe)

---

[2] The Initiative refers to the ongoing program of outreach and cooperation with six Mediterranean non-member "dialogue countries": Egypt, Israel, Jordan, Mauritania, Morocco, and Tunisia. See Ian O. Lesser, Jerrold Green, F. Stephen Larrabee, and Michele Zanini, *The Future of NATO's Mediterranean Initiative: Evolution and Next Steps*, MR-1164-SMD, 1999.

concern about the ever-increasing reach of missiles deployed on the European periphery, whether armed conventionally or with weapons of mass destruction (WMD).[3] Hard security risks also include spillovers of terrorism and political violence from conflicts in North Africa, the Balkans, and the Middle East. Some of these risks are of relatively greater concern to European allies. Many, especially missile proliferation, have direct implications for U.S. freedom of action.

Second, and significantly in light of changes in NATO and U.S. strategy, the Mediterranean region plays a critical role in power projection to the Middle East, the periphery of Europe itself (e.g., the Balkans), as well as the Maghreb and sub-Saharan Africa. The Mediterranean and Black Sea region is the logistical anteroom for power projection to the Gulf and the Caspian. Some 90 percent of the forces and materiel sent to the Gulf during operations Desert Shield and Desert Storm went by way of the Mediterranean. Yet assumptions about access, overflight, and transit (e.g., use of the Suez Canal) for extra-European operations cannot be taken for granted. Expeditionary approaches to presence and power projection, especially in relation to air power, strongly reinforce the importance of understanding and managing security relationships around the Mediterranean.

Third, the United States and its European allies share stakes in managing and coping with the consequences of specific crises on the southern periphery. In this respect, the area stretching from the Western Sahara to Central Asia and the Gulf, and the Mediterranean itself, contains an extraordinary number of flashpoints capable of imposing demands on Allied diplomacy and military power. Kosovo is only the latest example. Many of the most compelling problems for policymakers and planners on both sides of the Atlantic are to be found along this "arc of crisis." Most NATO planning contingencies are within the Southern Region, and the majority could involve Turkey in one way or another. This broad area also offers critical opportunities for foreign and security policy, from a Cyprus settlement to the Middle East peace process, from rethinking relations with Iran to the development and

---

[3] See Ian Lesser and Ashley Tellis, *Strategic Exposure: Proliferation Around the Mediterranean*, RAND, MR-742-A, 1996.

transport of Caspian energy resources. All have the potential to affect European security and America's role as a global power.

The geopolitics of NATO's southern periphery at the opening of the 21st century suggest a future dominated by security challenges that cut across traditional regional lines. European, Middle Eastern, and Eurasian security will be increasingly interwoven, with implications for the nature of risks facing the Alliance. The transregional character of the strategic environment will also imply new directions for strategy and the employment of military instruments. The United States and the U.S. Air Force will have to work with allies and others across the region in new ways, reflecting changing security agendas and strategies.

A NOTE ON KOSOVO

The research for this report was completed before the Kosovo crisis unfolded in the spring of 1999. In revising the study for publication, the Kosovo experience is acknowledged where relevant. A full analysis of the implications of the crisis will be undertaken in future RAND research. In most instances, the Kosovo experience strongly reinforces the findings of this report.

STRUCTURE OF THE ANALYSIS

This study analyzes the changes and new challenges in NATO's south, and their meaning for U.S. strategy and the U.S. Air Force. Chapter Two discusses the changing significance of the southern periphery for European security and U.S. strategy, including changes in the character of NATO's Southern Region itself and the meaning for bilateral relationships. Chapter Three explores the emerging transregional security environment in its political, economic, and military dimensions, with an emphasis on implications for U.S. and NATO power projection. Chapter Four examines issues concerning Turkey and security in the eastern Mediterranean. Chapter Five discusses the Mediterranean dimension of NATO adaptation, including Southern Region perspectives and the outlook for Alliance strategy. Finally, Chapter Six offers conclusions and implications for U.S. and NATO policy, and for Air Force planning, including areas for future research.

Chapter 2

The Southern Periphery and European Security

THE END OF MARGINALIZATION?

To understand the emerging security environment in the Mediterranean, it is useful to recognize the transformation that has taken
place over the last decades.[1] The Cold War made an early appearance
on Europe's Mediterranean periphery with the enunciation of the Tru-
man doctrine and the commitment to oppose Soviet destabilization of
Greece and Turkey.[2] In other respects, the south played a marginal role
in the East-West strategic competition and NATO strategy. Several factors contributed to the marginalization of what in Cold War parlance
was referred to as the "southern flank." First, the locus of threat was
genuinely in the center of Europe, and specifically against the territorial integrity of West Germany. The great debates about NATO's nuclear
and conventional strategy, transatlantic "coupling," and arms control
all concerned, first and foremost, the security of western Europe in the
face of a potent Warsaw Pact military threat. In theory, NATO Article
V commitments applied equally to the defense of all members. In reality, the defense of Hamburg and the defense of Athens were never
equivalent concerns for the Alliance.

Second, security concerns on the southern periphery were closely
tied to assumptions about the likely character and duration of an
East-West conflict. The Alliance did have points of exposure on its
southern flank—in northern Italy, Thrace, and the Caucasus, and in relation to the sea and air lines of communication stretching from the

---

[1] A good survey of changing Southern Region and other security perspectives on the Mediterranean can
be found in "Western Approaches to the Mediterranean," *Mediterranean Politics*, Special Issue, Vol. 1,
No. 2, Autumn 1996. See also Ian O. Lesser, *Mediterranean Security: New Perspectives and Implications
for U.S. Policy*, RAND, R-4178-AF, 1992.

[2] For a comprehensive review, see Bruce R. Kuniholm, *The Origins of the Cold War in the Near East:
Great Power Conflict and Diplomacy in Iran, Turkey, and Greece*, Princeton University Press, Princeton,
NJ, 1980.

Azores to Suez and the Black Sea. But these would only be pressing defense concerns for NATO planners in the context of a longer, conventional war (risks to Persian Gulf oil were also part of this equation) in which "theater interdependence" could be a real factor in the outcome. Most of the Southern Region, and certainly the western basin of the Mediterranean, was an area of relatively low risk and diffuse interest in Alliance affairs. The United States, through its bilateral defense relationships and its air and naval presence, was the key unifying element in the strategic equation. The transatlantic link was of particular importance to southern Europe and Turkey throughout the Cold War because of the complex problem of strategic coupling as seen from the south. Here, the problem was not only to ensure the U.S. commitment to European security, but also to link Southern Region security to the central concerns of Alliance decisionmakers.

Third, the political-military atmosphere in the south was further complicated by the residue of decolonization and frictions between Arab nationalist regimes and Europe, especially France. Indeed, if not for the controversy over French policy, pre-independence Algeria might well have fallen within the NATO area of responsibility in the 1950s. Contemporary arguments about the Mediterranean as a potential area of confrontation between Islam and the West, reflecting a very old concern, were anticipated by relations after 1945 in which nationalism rather than Islam was the motivating factor. The disengagement of France from mainstream NATO affairs also contributed to the marginalization of the south. Full French participation in the Alliance might well have given greater weight to the Mediterranean, Africa, and the Middle East, where French interests are heavily engaged, and where French military capabilities are suited to regional intervention. It is revealing that despite France's arm's-length approach to the Alliance, it is in the Mediterranean that French military cooperation with NATO and the United States has been most wide-ranging and effective.

Beyond the competition with the Soviet Union, the security environment in the south was relatively benign through the end of the Cold War. In the early 1980s, Balkan instability was not a concern. Arab-Israeli and Greek-Turkish frictions were dangerous regional problems, but unlikely to pose direct threats to western Europe. Developments in North Africa and elsewhere were, for the most part, not yet seen through a civilizational lens. Middle Eastern terrorism did manifest itself in Europe, and in the wake of two oil crises, much attention was paid to energy security. But Caspian oil was not yet on the agenda, and gas imports from North Africa were limited. The Iran-Iraq war pointed to the potential for missile attacks in regional conflicts, but observers were far from concerned about the implications for the security of NATO members.

In sum, the Cold War left a legacy of military and political marginalization within NATO's Southern Region, and with regard to the Mediterranean in general. NATO's southern members tended to be underrepresented in NATO commands, and defense infrastructure in the Southern Region remained undercapitalized on both a national and Alliance basis. Most significantly, NATO and its leading member states spent relatively little time and effort on problems of strategy in the south.

A Transformed Southern Region

Today's Southern Region is substantially transformed in political and military terms, with significant implications for NATO strategy and for the United States as a European and global power. Southern Europe is more active, more capable in military terms, and more central to Alliance strategy. At the same time, centrifugal trends are at work, especially in the eastern Mediterranean where Turkey is increasingly active.

Politically, the southern European landscape has been transformed over the past decades by the consolidation of democratic transitions in Portugal, Spain, and Greece. Much of the ambivalence about relations with the United States and NATO—the result of historical associations between Washington and previous totalitarian regimes—has also waned, especially in Athens, and in Madrid, where full integration in NATO has been a strong interest of recent socialist and conservative governments.[3] The consolidation of the U.S. military presence in these

---

[3] This change of attitude is most pronounced in Greece, despite the public opposition to NATO policy in Kosovo. A decade ago, NATO action against Serbia might have precipitated a break with the Alliance.

countries and the closure of bases (such as Torrejon near Madrid and Hellenikon outside Athens) have considerably eased the public acceptance climate. Issues surrounding the U.S. use of Lajes air base in the Azores may still be an important part of U.S.-Portuguese relations, but such issues are now placed in a broader frame by both sides.

Portugal, Spain, Italy, and Greece are now part of the European mainstream in their approach to defense issues.[4] Southern European states are in the process of streamlining their military establishments with a view to fielding smaller, more mobile forces, capable of participating in allied power projection missions.[5] In the case of Spain and Italy, this process is yielding forces with some capability for regional (i.e., Mediterranean) interventions. For the smaller NATO members, especially in the south, the ability to place defense requirements in a multilateral context—transatlantic or European—has emerged as a political necessity. It is notable that the trend toward Europeanization affecting southern European countries has not extended to Ankara, where political trends have set Turkey apart, and where high levels of defense spending and growing activism in security policy have very different sources. This issue and its implications are treated in detail in Chapter Four.

Bilateral defense relationships across the Southern Region have matured with the decline of traditional security assistance. This transformation began in the western Mediterranean and has recently been completed with the end of all grant assistance for Greece and Turkey. Southern Region states, especially Greece and Turkey, have also been recipients of equipment transferred under the Southern Region Amendment (a U.S. congressional measure) or "cascaded" from the United States and Germany in the early 1990s as a result of Conventional Forces in Europe (CFE) treaty-mandated reductions. The end of such transfers has removed security assistance from bilateral political agen-

---

[4] Much of this process of Europeanization has been fueled by substantial increases in southern European prosperity over the last decade. Edmund L. Andrews, "Europe's Clunkers Shift to the Fast Lane," *New York Times*, July 9, 1998.

[5] Southern European participation in the Combined Amphibious Force Mediterranean (CAFMED), part of NATO's STRIKFORSOUTH (Striking Forces Southern Europe), is one manifestation of this new capability. Paolo Valpolini, "Mediterranean Partnership for NATO Amphibious Forces," *Jane's International Defense Review*, July 1, 1998, p. 28.

das, but has also cast issues surrounding commercial arms transfers in sharper relief. In addition, the diversification of Southern Region defense-industrial and training links has also affected military-to-military relationships at the bilateral level. Whereas air forces in southern Europe and Turkey have traditionally been shaped by technical and training relationships with the USAF, new generations of Southern Region officers are as likely to train and fly with European counterparts. To the extent that such ties assist in creating a favorable climate for bilateral cooperation (e.g., on access issues), these changes suggest the need for new activities designed to reinvigorate military-to-military engagement.[6]

THREE SEMINAL CRISES

The Gulf War

Even more than the end of the Cold War, the Gulf War was a milestone in the evolution of Mediterranean security and the role of NATO Southern Region countries. Desert Shield and Desert Storm were not formally NATO operations, but Alliance planning, procedures, and habits of cooperation played an important role in coalition activity.[7] The Gulf War highlighted the preeminence of the Mediterranean for power projection further afield. Operations in the Gulf were heavily dependent on the logistical link stretching from the Atlantic to the Indian Ocean. Some 90 percent of the materiel required to support coalition operations in the Gulf arrived via the Mediterranean.[8] If airlift through southern European and Mediterranean air space is taken into account, this figure is undoubtedly even higher. The Gulf War and subsequent crises in the region also highlighted the significance of access to the Suez Canal as a means of shifting forces between theaters.[9] Denial or con-

---

[6] This point was emphasized by interlocutors in Spain, but applies elsewhere across the region.

[7] See Jonathan T. Howe, "NATO and the Gulf Crisis," *Survival*, Vol. 33, No. 3, May/June 1991.

[8] AFSOUTH estimate, cited in North Atlantic Assembly, "Draft Interim Report of the Sub-Committee on the Southern Region," 1991, p. 10.

[9] See Douglas Menarchik, *Powerlift—Getting to Desert Storm: Strategic Transportation and Strategy in the New World Order*, Praeger, Westport, CT, 1993.

straints on U.S. access to the Canal would severely complicate planning for Gulf contingencies, and might enormously magnify demands on airlift, access, and overflight, as well as demands on diplomacy with regional allies.

The connection between Mediterranean and Gulf security was operational as well as logistical. The eastern Mediterranean is closer to Baghdad than the southern Persian Gulf, and sorties from Incirlik in southern Turkey played a major role in the air campaign against Iraq. Nor was Turkey the only Southern Region country to offer its facilities for offensive air operations. Spain allowed B-52 sorties from Moron, despite public acceptance concerns.[10] Portugal, Spain, Italy, and Greece all sent naval forces to the Gulf, and allowed extensive use of their facilities for logistical purposes.

This degree of Southern Region cooperation was remarkable, especially against the background of historical ambivalence about "out-of-area" defense cooperation with the United States (only Portugal cooperated in the 1973 resupply of Israel, and no Southern Region member was willing to offer facilities in support of the 1986 El Dorado Canyon operation against Libya).[11] It can be explained, in part, by a softening of attitudes toward security cooperation among Spanish and Greek leaderships. It is likely that the progressive Europeanization of southern European defense policies, noted earlier, also played a role. Madrid and Athens were able to contribute precisely because there was a European consensus to do so. This "Brussels factor" is very much part of the post–Gulf War equation, and will play a role in the future calculus of cooperation between the United States and its southern European allies. Turkey, an even more significant actor in facilitating Western power projection beyond the Mediterranean basin, has not been part of this trend but could be more heavily affected by it in the future. Where a European consensus on cooperation has been absent, as in subsequent confrontations with Iraq, securing the cooperation of Southern Region states for access and overflight has proven to be difficult.

---

[10] Spain supported some 5000 sorties by U.S. aircraft during the Gulf War.

[11] The political and logistical complications of the air resupply to Israel are discussed in David R. Mets, *Land-Based Air Power in Third World Crises*, Air University Press, Maxwell Air Force Base, AL, July 1986, pp. 105–108.

The Gulf experience affected Southern Region security perceptions in other ways. It reinforced the interest in refashioning force structures for multinational operations beyond the NATO area. From Madrid to Athens, the Gulf crisis saw military leaderships pressing their more reluctant political counterparts to authorize additional contributions to coalition operations. (In Ankara, the situation was reversed, with a forward-leaning Ozal government committing Turkey to supportive policies over the reservations of a conservative military leadership.) The conduct of operations in the Gulf also made clear to defense planners that Southern Region militaries were, on the whole, ill prepared to wage modern, firepower-intensive and mobile warfare. The operational lessons of the Gulf War were taken most seriously by allies in the eastern Mediterranean facing tangible military threats. The Scud missile attacks on Israel and Saudi Arabia (and the exaggerated fears of possible Iraqi missile deployments in Mauritania threatening NATO's south) foreshadowed serious concerns about WMD proliferation.

The Gulf crisis pointed to a changing constellation of actors in Mediterranean security. As noted earlier, the conflict strongly reinforced the role of NATO as a focal point for operations in the south. It highlighted the role of France as a Mediterranean power, and saw German forces in strength in the region for the first time in the post–Cold War period. During the Gulf crisis, much of the German surface fleet moved to the Mediterranean to release other NATO assets for operations in the Gulf (this was also the first time since 1945 that an American aircraft carrier was absent from the Mediterranean). Overall, the Gulf crisis made clear that European security and the future of NATO would, in the future, be more deeply affected by developments outside the traditional NATO area, and that the U.S. military presence in Europe would face increasing demands from extra-European crises.

Algeria

In a very different sense, the crisis in Algeria also put Mediterranean security questions on the transatlantic agenda. Since the cancellation of election results in Algeria in 1991 and the onset of large-scale violence, Europe—especially France, Spain, Portugal, and

Italy—has been focused on the implications of the crisis for security on both sides of the Mediterranean. Several concerns stand out apart from the scale of the violence itself, with some 100,000 or more killed on all sides. European analysts have been concerned about the potential for disastrous refugee flows, although to date, there has been little effect on the flow of legal and illegal migrants across the Mediterranean. The fear of a potential Algerian refugee crisis has also been part of a wider and highly politicized European debate about migration from North Africa. A second concern has centered on the activities of Algerian Islamists, their supporters within Europe, and the potential for spillovers of terrorism and political violence. The hijacking of an Air France jet in 1994, the wave of Algerian-related terrorism in Paris in 1995–1996, and the discovery of networks affiliated with the GIA (Armed Islamic Group) across western Europe fueled this concern.[12] The violence in Algeria has lost its overtly political character (i.e., Islamic radicals versus the secular regime) and has become more complex—with factional feuds, economic opportunism, and tribal vendettas playing an active role—with the result that fears of an Islamic takeover have been supplanted by fears about the consequences of long-term instability in Algeria. This third concern is made more concrete by the growing European dependence on Algerian natural gas.

Europe has made few political overtures in relation to the crisis, and the few initiatives that have emerged have been firmly rebuffed by the military-backed regime in Algiers. The Algerian crisis has, however, encouraged a more general European concern about the future of poor, unstable societies across the Mediterranean and the implications for prosperity and security in Europe. In the absence of the Algerian crisis, it is unlikely that the European Union's (EU's) Euro-Mediterranean Partnership (the "Barcelona Process") would have developed.

The crisis has also been a driving force behind the evolution of NATO's Mediterranean Initiative. From an American perspective, the transatlantic aspects of the Algerian crisis have been especially significant. During a period when U.S. policymakers have concentrated on

---

[12] Most recently, Belgium has been threatened with GIA-led terrorism over the prosecution of Algerian extremists, and there have been arrests of Algerians with alleged ties to the GIA in North America.

NATO enlargement and the consolidation of political and security changes in the east, France and southern Europe have made it clear that the situation in North Africa must also be taken into account. From the mid-1990s onward, the Western European Union (WEU) has engaged in initiatives and exercises with clear relevance to crisis management in North Africa, in particular the formation of EUROFOR (Rapid Deployment Force) and EUROMARFOR (European Maritime Force). European planners have also explored with their American counterparts the possibility of U.S. support for European-led operations in the western Mediterranean.

The Balkans

Bosnia and continuing crises in the Balkans, above all Kosovo, have had the effect of enlarging, at least in a de facto sense, the NATO area of responsibility, as well as the field of actors in crisis management to NATO's south. Balkan crises have shifted security attention southward and have engaged political and military establishments from both sides of the Mediterranean. Every NATO Southern Region state has participated in some fashion in IFOR (Peace Implementation Forces), SFOR (Peace Stabilization Forces), and KFOR (Kosovo Peacekeeping Force). Italy, Turkey, and Greece have played an active role on the ground and have achieved considerable success in restoring stability in Albania through Operation Alba (again, with a strong concern about refugee flows). Three nonmember states involved in NATO's Mediterranean Initiative—Egypt, Jordan, and Morocco—also participated in IFOR/SFOR.[13] Following on from the experience during the Gulf War, and as part of a generally more open attitude toward missions beyond territorial defense, Germany has emerged as an active participant in Balkan peacekeeping. To the extent that Germany continues to pursue a forward-leaning policy on military operations beyond its borders, the Mediterranean is the most natural and likely sphere for action. Finally, Bosnia, Albania, and Kosovo have brought attention to humanitarian intervention as a source of new demands on military forces around the Mediterranean.

---

[13] It is likely that these and other non-NATO Mediterranean states will also contribute to KFOR.

Chapter 3
The New Transregional Security Challenges

THE END OF GEOGRAPHY?

A key feature of the strategic environment in NATO's south is the extent to which traditionally separate European, Middle Eastern, and Eurasian security questions are increasingly interwoven.[1] Indeed, one of the difficulties that NATO and the United States have had in addressing Mediterranean security in the past has been the bureaucratic and intellectual difficulty of treating problems that cut across established security theaters. Today's environment is characterized by security problems that transcend regional divisions, as well as new transregional perceptions and alignments.[2] The result is not so much the "end of geography" but rather the enlargement of traditional conceptions of the European security space.[3] This trend can be seen in key geopolitical, economic, and defense developments around the Mediterranean basin.[4]

---

[1] For discussion of post–Cold War geopolitics on the European periphery, see Martin W. Lewis and Karen E. Wigen, *The Myth of Continents: A Critique of Metageography*, University of California Press, Berkeley, 1997; George J. Demko and William B. Wood (eds.), *Reordering the World: Geopolitical Perspectives on the 21st Century* Westview, Boulder, CO, 1994; Peter J. Taylor, "From Heartland to Hegemony: Changing the World in Political Geography," *Geoforum*, Vol. 25, No. 4, 1994; and Michael C. Desh, "The Keys That Lock Up the World: Identifying American Interests in the Periphery," *International Security*, Summer 1989.

[2] The North Atlantic Assembly has been particularly attuned to these trends. See reports of the Mediterranean Special Group, including Pedro Moya, "NATO's Role in the Mediterranean," North Atlantic Assembly (NAA), Brussels, October 1997.

[3] I am grateful to Alvaro Vasconcelos of the Portuguese Institute for International and Strategic Studies for this intriguing formulation. His reference point was the changed role of the Azores in Western strategy as a result of technological and political developments.

[4] A comprehensive treatment of risk factors in the Mediterranean can be found in *NATO's Southern Flank*, Third Report of the Defence Committee, House of Commons, The Stationery Office, London, 1996.

THE GEOPOLITICAL DIMENSION

The end of the Cold War and developments in Europe, North Africa, and the Middle East have encouraged new linkages between regions around the Mediterranean basin. Perceptions of the United States and NATO in the Islamic and Arab world have been strongly influenced by developments in Europe, especially in the Balkans where Muslim affinities are engaged. Governments and publics from North Africa to Turkey were critical of perceived European inaction in Bosnia. Russian actions in Chechnya raised similar concerns. NATO policy in Kosovo has been more positively received, although tempered with ambivalence about the potential for Western intervention elsewhere.

The status of Muslim and Arab immigrants in western Europe is another issue that commands attention on Europe's periphery. States such as Morocco, Algeria, and even Turkey have come to view the treatment of their nationals abroad as part of their own diplomatic agenda—an issue that has acquired greater significance with the rise of anti-immigrant violence by far-right groups in France, Italy, Germany, and elsewhere.[5]

Lack of progress in the Middle East peace process, and Arab frustration with the direction of European and especially U.S. policy in this regard, has severely constrained north-south cooperation on security issues around the Mediterranean since the mid-1990s. The status of the peace process has made it difficult to engage Arab partners in NATO's Mediterranean Initiative, particularly multilateral confidence-building measures that would include Israel. It has also limited the EU's ability to develop the security aspects of the Euro-Mediterranean partnership. The May 1999 Israeli elections could transform this situation and open new avenues for security cooperation in the context of the peace process. Europe, for its part, worries about the spillover effect in Europe of political struggles in the Middle East. Algeria and the Kurdish problem provide leading examples.

In the eastern Mediterranean, new geopolitical alignments are changing the strategic environment, as well as the context for U.S. and allied power projection. The most significant development has been the

---

[5] Spain also experienced a wave of violence against North Africans in 1999. See "Spain and Race: Trouble," *The Economist*, July 24, 1999, p. 47.

burgeoning relationship between Turkey and Israel, and the potential for a wider Israeli-Turkish-Jordanian-U.S. alignment. The Israeli-Turkish relationship is multidimensional and evidently "strategic," although both countries have been careful not to overstress this aspect. Politically, recent agreements codify a relationship that has existed in a less visible fashion for decades. Economically, the way is now open to greater cooperation in many areas, including joint ventures in Central Asia, the energy sector, and water. Militarily, the range of cooperative activity is large and spans intelligence cooperation, joint training and exercises, and defense-industrial projects. Air power cooperation has been a particular feature of the relationship, with Israeli aircraft gaining access to Turkish air space and a variety of ongoing and potential aerospace projects.[6] Strategically, the relationship serves to pressure Syria—a common concern—and to provide reassurance against Iraqi and Iranian risks, especially in relation to WMD and terrorism.

Jordan, with its limited defense capabilities but shared concern about powerful Arab neighbors, is a much more ambivalent third partner in this new alignment.[7] Limited air and ground force exchanges have already occurred between Turkey and Syria, and Jordanian observers have participated in trilateral Israeli-Turkish-U.S. naval exercises.[8] These developments have, in turn, provoked a critical reaction from Arab states in the region, above all Syria and Egypt. In response, relations between Syria and Iraq, historically difficult, have grown closer. There has also been considerable speculation about closer security cooperation among Syria, Greece, and Cyprus as a (not very promising) counterweight to Turkish cooperation with Israel.[9] To this might be added the possibility of Russian support, following on from the supply of military equipment to Cyprus.

---

[6] These include Israeli upgrades to Turkish F-4s and F-5s, and likely Turkish involvement in Israel's Arrow ballistic-missile defense program.

[7] Turkey and Jordan have exchanged infantry companies and have begun to collaborate on exchanges of aircraft. The Prime Ministers of Jordan and Turkey have called for "comprehensive" security cooperation in the eastern Mediterranean and the Middle East. See "Jordan and Turkey Call for Security Plan," *International Herald Tribune*, September 7, 1998.

[8] Jordan has announced that it will not participate in the next U.S.-Turkish-Israeli naval exercises.

[9] The existing Greek-Syrian defense cooperation agreement does not appear to be as ambitious as some Turkish analysts have suggested. In any event, it is doubtful that the Greek air force is equipped to exploit advantages that might flow from access to Syrian bases in a crisis with Turkey over Cyprus or the Aegean. There have also been reports of Greek defense agreements with Iran and Armenia; their content, if any, is unclear.

These new "geometries" are encouraging a greater degree of interdependence in security terms among Europe, Eurasia, and the Middle East. They also present new opportunities for U.S. strategy, particularly in relation to the projection of air power. Political acceptance and force protection problems may complicate the longer-term outlook for forward presence in the Gulf. The southern route for power projection from the Indian Ocean faces similar access constraints, and depends heavily on the predictable use of the Suez Canal. In this setting, an alternative northern approach to power projection for the Gulf may be attractive, logistically as well as politically. Baghdad is closer to the eastern Mediterranean coast than it is to the southern Persian Gulf (roughly 500 versus 1000 miles). A northern strategy for the Gulf would imply a more active role for Israel, Jordan, and Turkey as partners in expeditionary operations beyond the eastern Mediterranean. To be sure, formal, standing arrangements for U.S. access and overflight would face formidable political obstacles. Progress in the peace process—or other dramatic developments in the region, such as a conflict between Turkey and Iran, could change this equation.[10]

THE ECONOMIC DIMENSION AND ENERGY SECURITY

Europe is the key economic partner for all of the southern Mediterranean states, especially in the energy sector. Much EU attention and funding (although far less than that given to central and eastern Europe) is devoted to encouraging development and investment in non-member Mediterranean states.[11] The EU's Euro-Mediterranean Partnership (the "Barcelona process") envisions the establishment of a comprehensive Mediterranean free trade area by 2015. The EU has also negotiated a series of association agreements around the Mediterranean, of which the customs union agreement with Turkey is the most ambitious.[12] The poor state of Turkish-EU relations in the wake

---

[10] See Zalmay Khalilzad et al., *The Implications of the Possible End of the Arab-Israeli Conflict for Gulf Security*, RAND, MR-822-AF, 1997.

[11] The assistance to be given as part of the Barcelona process is a small fraction of the EU "cohesion funds" allocated to southern European members.

[12] Other agreements have been concluded with Morocco, Tunisia, the Palestinian Authority, and Jordan. Agreements are pending with Algeria, Lebanon, Syria, and Egypt. The EU plans to open accession talks with Cyprus and, quite possibly, Malta.

of the Luxembourg, Cardiff, and Cologne summits was all the more disturbing because the relationship with Ankara is, in many respects, the cornerstone of the EU's Mediterranean strategy. Although the EU's Helsinki summit decision to include Turkey in the list of candidates for membership has put relations back on track, the long-term outlook for Tuskish integration remains uncertain. The EU approach to the southern Mediterranean as a whole may be characterized—not unreasonably—as an attempt to subsidize political stability and dampen migration pressures in the south.

Energy and energy security are key transregional issues from Gibraltar to the Gulf and the Caspian. Protecting access to the energy resources of the Persian Gulf has imposed power projection requirements on the United States and on at least some NATO allies that have, in turn, shaped security relationships around the Mediterranean. More directly, the eastern Mediterranean has been a terminus for Turkish and Syrian pipelines, providing an alternative to shipment through the Gulf and the Strait of Hormuz. Suez, although unsuitable for the largest crude carriers, also plays a role in the transport of Gulf oil to European markets.

Since the break-up of the Soviet Union, the prospect of oil and gas supplies from the Caspian Basin has emerged as a new factor affecting security perceptions and planning.[13] The Byzantine politics of oil pipeline alternatives and the likelihood that choices will ultimately be driven more by private, economic decisionmaking than by geopolitical design complicate informed discussion of the security implications. Despite U.S. and Turkish interest, and commitments in principle, the outlook for construction of the Baku-Ceyhan pipeline across Turkey to a terminal on the eastern Mediterranean is uncertain. A route from Baku to Supsa on the Black Sea is more likely, with all that this implies for increased tanker flow through the Bosporus. Eventually, changes in Iranian relations with the West may result in much Caspian oil passing through Iran to the Gulf (increasing the dependence on unimpeded passage through the Strait of Hormuz). The consensus view suggests that Caspian resources will provide an important new long-term source of

---

[13] See Richard Sokolsky and Tanya Charlick-Paley, *NATO and Caspian Security: A Mission Too Far?* RAND, MR-1074-AF, 1999.

energy for world markets, although still less significant than other Middle Eastern sources and far from the transforming development that early analyses implied.[14]

Nonetheless, the development of Caspian resources and lines of communication to world markets will have the effect of tying Europe, Eurasia, and the Middle East more closely together in security terms. This tendency will be reinforced by the parallel development of new overland nonenergy lines of communication. New road and rail links, and the improvement of Black Sea ports, will make it possible for Caucasian and Central Asia states to bypass Russia in their trade with Europe and the Middle East.[15] Regardless of whether Caspian oil itself emerges as a strategic stake warranting Western planning for its defense, the development of new resources in the region is likely to offer more opportunities for geopolitical friction touching on the interests of NATO allies. Turkey, adjacent to some 70 percent of the world's proven oil reserves and some 40 percent of world gas reserves, will be in a position to affect—and be affected by—developments within the "strategic energy ellipse" formed by the Caspian and the Gulf.[16]

The longstanding interest in Gulf oil and the fashionable interest in Caspian "great games" have obscured attention to another and perhaps equally important development in energy geopolitics on Europe's periphery—growing European reliance on imported natural gas. Europe depends on North Africa for roughly 25 percent of its natural gas requirements. For southern European countries and France, the dependency is far higher. Algeria is the leading and increasingly influential supplier. Spain already depends on Algeria for some 70 percent of its supply, and this figure is set to increase substantially over the next decade as a result of fuel choices and new lines of communication for gas. Portugal is almost entirely reliant on Algerian gas supplies.[17] Most of this supply reaches Europe through two pipelines—the Trans-Med system that connects North Africa with Italy and the new Trans-

---

[14] See unclassified CIA estimates cited in Sokolsky and Charlick-Paley.

[15] See Abraham S. Becker, *Russia and Caspian Oil: Moscow Loses Control*, RAND, P-8022, 1998.

[16] Geoffrey Kemp and Robert E. Harkavy, *Strategic Geography and the Changing Middle East*, Carnegie/Brookings Institution Press, Washington, DC, 1997, pp. 109–153.

[17] Gas import estimates compiled by RAND colleagues Nurith Bernstein and Richard Sokolsky from the Organization for Economic Cooperation and Development (OECD) and other sources.

Maghreb pipeline that supplies Algerian gas to Spain and Portugal (as well as France, Belgium, and Germany) via Morocco. Additional pipelines are planned to expand the Libya-Italy link and to provide Libyan gas to Egypt, and through Israel with links to Qatari gas, onward to Turkey.[18] New pipelines from the Caucasus and Central Asia, perhaps via Iran, will eventually provide gas to eastern and central Europe. As a result, by the early 21st century, "Europe will be profoundly tied into the Mediterranean region by its dependence on energy supplies through expensive fixed delivery infrastructure."[19] Moreover, unlike oil, the market for gas is regional, not global, and relatively inflexible in responding to specific interruptions.

At least one prominent observer has suggested that this trend makes the Middle East of "acute and growing interest to Europe for reasons that have nothing to do with American priorities"[20] The United States may, however, find itself sharing this energy security concern as a matter of Alliance interest and NATO strategy.[21] There are also certain parallels with the growth of European dependence on Soviet gas in the 1980s, a development that disturbed many U.S. strategists but which most Europeans saw as useful and stabilizing interdependence. In the case of North African gas, the significant risks to access are likely to come from turmoil and anarchy rather than deliberate cutoffs by suppliers or transit states. With growing levels of European dependence on gas from the southern periphery, it would not be surprising if NATO in the 21st century is compelled to plan for operations to restore the flow of gas from far-flung and unstable regions. Such operations are every bit as likely as contingencies involving Caspian oil, and far more likely to find broad support within the Alliance.

---

[18] George Joffé, "The Euro-Mediterranean Partnership: Two Years After Barcelona," Middle East Programme *Briefing*, No. 44, May 1998, p. 2.

[19] Joffé, "Euro-Mediterranean Partnership," p. 2.

[20] Ibid.

[21] NATO's new Strategic Concept refers to "disruption of the flow of vital resources" as one of several new challenges. See NATO press release NAL-S(99)65, paragraph 24.

THE DEFENSE DIMENSION

Two prominent security challenges emanating from Europe's southern periphery—proliferation and terrorism—are emblematic of the emerging transregional environment. Some additional, "soft" security problems—refugees, drugs, and crime—are also worth noting in this context.

Despite some resistance to debate on this topic in NATO circles, analysts now recognize that the Southern Region is increasingly exposed to proliferation risks as a result of the growing reach of ballistic missile systems deployed in the Middle East and potentially deployable in North Africa and elsewhere around the Mediterranean. Cruise missile risks have received less attention in this setting, but probably should be taken more seriously in the Southern Region.[22] Turkish population centers are already within reach of missiles deployed by Iran, Iraq, and Syria. Libya or Algeria could acquire systems from diverse suppliers (e.g., China, North Korea, and Pakistan) that could easily reach southern European capitals.[23] The 1998 test of a medium-range ballistic missile by Iran, and planned tests of a new multistage missile with a 2600-mile plus range, point to the possibility that, within the next decade, not only Southern Region capitals but Paris, London, Berlin (and possibly Moscow) will be within range of systems deployed on Europe's periphery. The attention to proliferation risks within NATO, especially in NATO's south, has been sporadic but is growing since the Gulf War. The prospect of much broader, Alliance-wide vulnerability to these risks suggests that this challenge is set to receive more serious attention in the future.

The leading motivations for the proliferation of weapons of mass destruction and the means for their delivery at longer ranges are, arguably, regional rather than transregional—that is, largely south-south rather than north-south. The most likely victims of WMD and ballistic missile use are in the south, as the Iran-Iraq war, the use of Scud missiles in Yemen, and even the Gulf experience suggest. But the prolifer-

---

[22] See Dennis M. Gormley, "Hedging Against the Cruise Missile Threat," *Survival*, Vol. 40, No. 1, Spring 1998.

[23] See Lesser and Tellis, *Strategic Exposure*. For a more recent analysis, see Thanos Dokos, *Proliferation of Weapons of Mass Destruction and the Threat to NATO's Southern Flank: An Assessment of Options*, NATO Research Fellowship, Final Report, June 1998.

ation of longer-range systems, even conventionally armed, can change the dynamic of north-south security relations on Europe's periphery in significant ways, and with transatlantic implications. In the broadest sense, growing missile reach means that Europe (and U.S. military facilities in Europe) will be exposed to the retaliatory consequences of Western action outside of Europe, whether in the Gulf, North Africa, or elsewhere. Europe will also be more directly exposed to political blackmail in peripheral crises, energy disputes, or other conflicts along north-south lines. Nor will the potential for blackmail be limited to the Middle East. "Rogue" states in the Balkans or the former Soviet Union may be in a position to threaten European territory and constrain European policy.[24]

Missile proliferation in the Middle East can affect European security in another, less-direct fashion. Many of the world's leading WMD and missile proliferators are arrayed along an arc from Libya to South Asia. As these states acquire more destructive capacity and advanced delivery systems, there is the potential for allies such as Turkey to respond by developing deterrent capabilities of their own. The prospects for such reaction are reduced if there is sufficient confidence in the NATO security guarantee, including its nuclear dimension. But under certain conditions (including a marked increase in tensions with Syria, Iraq, or Iran), Ankara might well decide to develop sovereign capabilities. The strategic dilemma would be even more pronounced in the event of a nuclear Iran or Iraq. Proliferation pressure on Turkey could in turn fuel a missile race in the Balkans and the Aegean, threatening already unstable relationships in southeastern Europe.

U.S. freedom of action across the Mediterranean and in Europe could also be affected by proliferation trends. Increased European security exposure to Middle Eastern crises may have operational as well as diplomatic consequences. A more vulnerable Europe will demand a more active role in Middle Eastern diplomacy, and in ways that could affect NATO's ability to address extra-European problems. At the same time, the potential exposure of European population centers to retaliation could complicate the prospects for U.S. access to southern European and Turkish facilities for expeditionary operations in North

---

24 This concern has been raised in relation to Serbia with its rudimentary missile capability.

Africa and the Middle East—and perhaps even in the Balkans. Under these conditions, the United States and its allies might not enjoy the luxury of a secure rear area from which to project military power outside Europe. The prospects for access might then depend, in part, on the ability to provide a reasonably effective area defense against missile attack.

Leaving aside the technical problems of missile defense in the Southern Region, efforts to address missile risks around the Mediterranean face significant obstacles. To the extent that NATO concentrates more heavily on proliferation challenges, these obstacles will likely loom larger, especially from the perspective of AFSOUTH. First, Southern Region allies tend to assume that the United States will eventually provide some form of mobile theater missile defense (TMD) for the region, either afloat or readily deployable by air. The costliness of such systems deters most southern allies from participation in the development of theater ballistic missile defense (TBMD) systems [Italy through the Medium Extended Air Defense System (MEADS) and possibly Turkey through the Israeli Arrow program are partial exceptions]. Second, the relative lack of modern communications links in the Southern Region severely complicates the task of addressing missile and WMD risks, especially in relation to warning time and civil defense. Third, Southern Region allies—with the exception of Turkey—are uncomfortable with counterproliferation strategies, preferring political approaches to proliferation dynamics, which they regard, with some reason, as largely south-south in nature.

If WMD and missile proliferation are the dominant concerns at one end of the "hard" security spectrum, terrorism and spillovers of political violence are equally prominent at the low-intensity end.[25] Political struggles and anarchic violence in Algeria have already produced spillovers of terrorism in France and Belgium, and officials on both sides of the Atlantic are concerned about the potential for wider terrorism carried out by networks sympathetic to the Armed Islamic Group (GIA) or simply by alienated North African immigrants. The Kurdish issue, including its violent component, has also been imported

---

[25] A good analysis of terrorism and other risks in the new European environment can be found in Alessandro Politi, "European Security: The New Transnational Risks," *Chaillot Paper*, No. 29, Paris WEU Institute for Security Studies, October 1997.

into Europe through the activities of the Kurdistan Workers Party (PKK). Germany, with some two million Turkish residents, at least a third of whom are Kurds, has been particularly exposed to spillovers from this ongoing struggle. Italy and Greece were forced to contend with the diplomatic embarrassment of having the PKK leader, Abdullah Ocalan, as an asylum seeker, giving rise to friction in relations with Ankara. The capture of Ocalan and the ensuing Kurdish protests across Europe have put the Kurdish issue and its transnational character in sharp relief. There is also an important nexus among Mediterranean terrorist activities, drug trafficking, and transnational crime. The PKK is a leading example in this regard.[26]

Conflicts in the southern Balkans and the possibility of further anarchy in Albania, Kosovo, or Macedonia, can spill over into NATO Europe to affect the security of Greece and Italy. Both countries would be profoundly affected by the arms smuggling, crime, refugee movements, and potential terrorism associated with zones of chaos across the Adriatic. Over the last decade, Athens and Rome have had to confront the political, social, and internal security problems posed by Balkan and Kurdish refugees. In the same period, Turkey has had to address the arrival of roughly 100,000 ethnic Turks from Bulgaria and could face disastrous new flows as a result of further Balkan upheavals or conflict in northern Iraq. Spain and Portugal confront on a smaller scale illegal migration from Morocco, but they worry that turmoil across the Mediterranean could produce more extensive movements of people.

These and other low-intensity, transregional risks are blurring the distinction between internal and external security in a European context. With the implementation of the EU's Schengen Agreement facilitating cross-border travel, southern Europe is in an increasingly uncomfortable position of responsibility for controlling the movement of people into the EU as a whole. There is a natural southern European interest in multilateralizing this problem, along with other cross-border challenges such as drug trafficking and international crime. Cooperation on such "third pillar" issues within EU institutions is one ap-

---

[26] See Ian O. Lesser, Bruce Hoffman, John Arquilla, David Ronfeldt, and Michele Zanini, *Countering the New Terrorism*, RAND, MR-989-AF, 1999.

proach. Changing NATO missions provide another opportunity for Southern Region countries to address security problems that the Alliance has traditionally considered national responsibilities.

The Mediterranean basin contains many crises and potential crises capable of producing conventional, cross-border conflicts.[27] Very few pose the risk of a direct military clash between north and south. Beyond the ongoing potential for conflict with "rogue" states such as Libya, only two scenarios stand out. In the western Mediterranean, there is a potential for conflict between Spain and Morocco over the Spanish enclaves of Ceuta and Melilla. Apart from participation in multinational peacekeeping operations, this remains the leading contingency for Spanish military planners and is technically outside the NATO treaty area. The outlook for this potential flashpoint depends critically on the character of post-Hassan Morocco. At the other end of the Mediterranean, Turkish-Syrian and Turkish-Iranian relations constitute flashpoints, with the potential for direct NATO involvement (this issue is taken up in the next chapter).

Overall, south-south risks predominate along Europe's southern periphery, but there are many possibilities for transregional spillovers. Europe, and the United States as a European power, will be exposed to the consequences of developments over the horizon—in North Africa, the Middle East, and Eurasia. Challenges emanating from the European periphery, from proliferation to migration, also suggest that the "new" NATO has important new consumers of security in its Southern Region. These security challenges are harder, more direct, and more likely to involve the use of force in the eastern Mediterranean, especially on Turkey's borders.

---

[27] A tour d'horizon would include potential conflicts between Spain and Morocco, Morocco and Algeria, Libya and Tunisia, Libya and Egypt, Israel and its Arab neighbors, Turkey and Syria/Iraq/Iran, Greece and Turkey, and the complex of rivalries and flashpoints in the Balkans.

Chapter 4

Turkey and Security in the Eastern Mediterranean

Turkey will be a critical partner in Alliance efforts to address security challenges on the European periphery, including risks associated with uncertain Russian futures.[1] The United States will have an independent interest in security cooperation with Turkey for power projection in adjacent areas of critical interest—the Balkans, the Caucasus and the Caspian, the Levant, and the Gulf. To the extent that NATO moves to become a more geographically expansive, power-projection alliance, this interest in Turkey's role will be more widely shared. But Turkey itself is experiencing profound internal change, and Turkey's relations with several of its neighbors remain troubled. Improved relations with the EU can have a positive effect in both dimensions.

INTERNAL UNCERTAINTIES

The future direction of Turkish external policy, and the future of Turkey as a security partner for the West will be driven to a great extent by internal developments. Even if the overall direction of Turkish policy remains steady and pro-Western, Turkey's ability to play an active role in adjoining regions and in NATO affairs (including the peaceful resolution of disputes with Greece) will depend on political stability in Ankara. The outlook is uncertain and is characterized by

---

[1] One recent analysis describes Turkey as a pivotal state in its own right. See Robert Chase, Emily Hill, and Paul Kennedy, "The Pivotal States," *Foreign Affairs*, January/February 1996. For a perspective from the early 1990s, see Graham E. Fuller, Ian O. Lesser et al., *Turkey's New Geopolitics: From the Balkans to Western China*, Westview, Boulder, CO, 1993. See also Andrew Mango, *Turkey: The Challenge of a New Role*, Praeger, Westport, CT, 1994; and Simon Mayall, *Turkey: Thwarted Ambition*, National Defense University, Institute for National Security Studies (NDU-INSS), Washington, DC, 1997.

flux on three broad fronts: secularism versus Islam, the state versus its opponents, and the future of Turkish nationalism.[2]

The Ataturkist tradition of statism, Western orientation, secularism, and non-intervention has been under strain for decades, critically so in the period since the Gulf War. The struggle between secular and Islamist visions of Turkey as a society has been a focal point for Turkish and Western observers since the electoral successes of the Refah Party and its leadership of a coalition government. The removal of Refah from power and the banning of the party and its leadership from Turkish politics thrust the military to the forefront. The 1999 general elections produced a nationalist coalition of the right and the left, with a sharp decline in support for centrist parties and for Refah's successor, the Virtue Party. The consolidation of military influence in defense of the secular state also means that, more than ever, the Turkish military is a key interlocutor on foreign and security policy issues.

Beyond the question of secularism versus Islam, the internal scene is defined by the broader conflict between the state and its opponents—from the religious right to the left. The defining struggle in this context is the ongoing war between Ankara and Kurdish separatists. Even apart from the capture of Abdullah Ocalan, the security forces have achieved considerable success in containing the PKK insurgency in the southeast, a conflict that has claimed perhaps 40,000 lives over the last decade. Cities in the southeast are now more secure, but the conflict is far from over and exerts a continuing drain on the Turkish economy and society. The human rights consequences of the war in the southeast and in northern Iraq have also imposed incalculable opportunity costs on Turkey in its relations with the EU and the West as a whole.

The improved security picture in the southeast is the result of better counterinsurgency techniques and aggressive cross-border operations in northern Iraq that have shifted much of the war against the PKK off Turkish territory. Yet little progress has been made on the

---

[2] For a provocative analysis on these lines, see Christopher de Bellaigue, "Turkey: Into the Abyss?" *The Washington Quarterly*, Summer 1998; for a series of more optimistic Turkish and foreign views, see "Turkey's Transformations," *Private View* (Istanbul), Autumn 1997. An excellent post-election analysis is offered in Alan Makovsky, "Ecevit's Turkey: Foreign and Domestic Prospects," *Policywatch* No. 398, Washington Institute for Near East Policy, July 20, 1999.

more fundamental issue of a political approach to Kurdish rights in the southeast (and for the majority of Kurds living elsewhere in Turkey). Without a resolution of this problem, there will remain some potential for Kurdish separatism and other forms of opposition to the state, including Islamism, to interact in unpredictable and potentially destabilizing ways. The social stresses resulting from a decade of strong economic growth and the rise of a dynamic private sector, but with increasing income gaps and high inflation, are other troubling elements on the internal scene. Corruption and a burgeoning illegal sector have further contributed to discredit the traditional political class and to inhibit the emergence of a credible centrist alternative in Turkish politics—an alternative that the Turkish military and business community, among others, would like to encourage.

Against this background of political and social change, Turkish nationalism has emerged as a powerful force uniting diverse elements—military and civilian, secular and religious. The strong showing of the Nationalist Action Party (MHP) suggests that Turkish nationalism has supplanted Islamism as a popular political force.[3] More vigorous nationalism is evident in closer attention to Turkish sovereign interests, greater affinity and activism in support of Turks abroad, and a more assertive and independent external policy. Rising Turkish nationalism, dating from the period of the Gulf War but gathering pace in recent years, has paralleled other important changes in Turkish foreign policy, notably the growing role of public opinion and the rise of potent lobbies (Turkish Cypriots, Bosnians, Azeris) in Ankara. These factors are now central to policymaking in key crisis areas, from the Aegean to the Caucasus, and in relations with the United States and Europe. Although Western policies, especially the attitude of the EU toward Turkish membership, are part of this new foreign policy equation, U.S. and European leverage is arguably more limited than in the past.[4]

---

[3] Despite its extremist past, the MHP garnered 18 percent of the vote in the 1999 elections, second only to the nationalist Democratic Left Party's 22 percent.

[4] On Turkish-EU Relations, see F. Stephen Larrabee, *The Troubled Partnership: Turkey and Europe*, RAND, P-8020, 1998; and Ambassador Ozdem Sanberk, "The Outlook for Relations Between Turkey and the European Community After the Cardiff Summit," remarks delivered at the Washington Institute for Near East Policy, July 20, 1998.

THE PRIMACY OF INTERNAL SECURITY CONCERNS

Turkish security perspectives are unique when viewed against the background of Alliance-wide concerns. No other member of the Alliance faces a similar range of external security challenges or such significant internal problems. Many of these challenges are typical of the post–Cold War security environment. At the same time, Ankara retains a high degree of concern about residual risks from Russian behavior. The net result is a degree of exposure and security consciousness unique within NATO.

Internal security concerns top the Turkish agenda and color perspectives on external actors such as Greece, Syria, and Iran. The Turkish General Staff consistently cites the struggle against antisecular forces (Islamists) and separatism (the PKK) as the number one and two defense priorities.[5] The former is largely a political and judicial sphere of activity. The latter consumes a good part of Turkey's defense resources and energy, and is seen as integral to guaranteeing the unitary character of the Turkish state.[6] Even in the midst of a large and costly defense modernization program (perhaps as much as $80 billion over the next decade, $150 billion over the next 25 years), the military devotes enormous resources to the conduct of operations against the PKK.[7] One consequence of this effort has been a steady improvement in the mobility and operational readiness of Turkish forces, a development with implications for the military balance with Syria, Iran, Greece, and Russia. For Turks, operations against the PKK are defined as counterterrorism, and the primacy of this activity gives Ankara a strong interest in seeing cooperation against terrorism incorporated in NATO discussions.

The battle against the PKK also provides the lens through which the Turkish military and civilian leadership view the situation in northern Iraq. The issue of northern Iraq is a leading source of suspicion in relations with the United States and Europe. The United States and Europe view northern Iraq as a function of policy toward Baghdad and

---

[5] These two priorities are sometimes reversed.

[6] Including within the military itself. There have been numerous purges of military officers for antisecular activities in recent years.

[7] Some 200,000 military and gendarmerie personnel are deployed for this purpose in the east and southeast of the country.

Iran, and are concerned about regional security and human rights. For Ankara, developments in northern Iraq are viewed in relation to their effect on Kurdish nationalism, separatism, and the viability of the PKK. Thus, Operation Provide Comfort was viewed with enormous skepticism by many Turks (although tolerated by the military), as are U.S.-brokered agreements between the two leading Kurdish factions. Both have been interpreted as policies aimed at fostering rather than dampening Kurdish aspirations at Turkish expense. Turkish policymakers have tolerated, but are clearly uncomfortable with, the use of Incirlik air base for strikes against Iraqi targets.[8]

Relations with Iran are also part of the Turkish security equation. There has been some concern among Turkish officials about an Iranian hand in Turkish politics, including philosophical and monetary support for Islamists. In all likelihood, the Iranian role in this regard has been minor. Sympathetic Turkish businessmen and, especially in earlier years, Saudi donations, have almost certainly played a larger role in the growth of Islamic institutions and politics in Turkey. Iran has also been implicated in support for the PKK, and reported Turkish strikes against PKK targets on Iranian territory have been part of a cyclical pattern in Turkish-Iranian tensions.[9] Iranian nuclear and ballistic missile programs are a source of growing concern in Ankara and reinforce the Turkish interest in intelligence and TBMD cooperation with Israel. These concerns also encourage a conservative view of NATO nuclear policy and a strong interest in counterproliferation as part of the new NATO agenda.

GREEK-TURKISH CONFLICT: OUTLOOK AND CONSEQUENCES

A third, more proximate, source of risk in Turkish perception concerns relations with Greece. Objectively, there can be little strategic rationale for premeditated conflict on either side. Open conflict would

---

[8] Between late December 1998 and July 1999, as part of Operation Northern Watch, U.S. and British aircraft operating from Incirlik struck Iraqi targets in the northern no-fly zone on more than 60 occasions. U.S. European Command (EUCOM) figures cited in *European Stars and Stripes*, July 27, 1999, p. 2.

[9] Gokalp Bayramli, "Tensions Heighten Between Tehran and Ankara," *RFE/RL Weekday Report*, July 20, 1999.

pose enormous political risks for both Ankara and Athens, quite apart from uncertainties at the operational level. Yet the risk of an accidental clash in the Aegean (on the pattern of the Imia-Kardak crisis of 1996) remains, given the continuing armed air and naval operations in close proximity and the highly charged atmosphere surrounding competing claims over air and sea space. NATO has achieved some success in convincing both sides to pursue military confidence-building measures agreed to in 1988, which might reduce the risk of incidents at sea and in the air. It is unclear whether these measures will be fully implemented.[10] On Cyprus, the planned delivery of Russian-supplied S-300 surface-to-air batteries and the Turkish threat to respond militarily to their deployment raised the stakes considerably.[11] The Cypriot decision not to deploy this system on the island defused an explosive situation.

The Greek-Turkish dispute has evolved considerably over the past decade, with significant implications for regional stability and crisis management. First, the geopolitical context has changed, with the end of Cold War imperatives giving both Greece and Turkey greater freedom of action. Turkey has become a more independent and assertive regional actor, although outside the European mainstream. Greece, by contrast, has become more European in orientation. Second, both countries have experienced substantial internal change over the past decade. Public opinion is a critical influence in both countries, reinforced by the growing role of the media in periods of crisis.[12] Nationalism may now be a more powerful force in Turkey, but for both countries, the Aegean and especially Cyprus are the nationalist questions *par excellence*. Third, there has been an increase in the scope of Greek-Turkish competition and the potential for linkage and escalation. Beyond traditional disputes over the Aegean and Cyprus and the status of minority communities in Thrace and Istanbul, developments in the Balkans and in the larger eastern Mediterranean balance (i.e., involv-

---

[10] See "Statement by the Secretary General of NATO, Dr. Javier Solana, on Confidence Building Measures Between Greece and Turkey," NATO Press Release, (98)74, June 4, 1998.

[11] See "Missiles and the Eastern Mediterranean: A Dangerous Game of Brinkmanship," *IISS Strategic Comments*, June 1998.

[12] Many Turkish and Greek observers point to the inflammatory role of the television media in both countries during the Imia-Kardak crisis.

ing Israel, Syria, and even Russia) are now part of the equation.[13] Greek strategists talk of Muslim "encirclement" in the Balkans. Turks allude to a looming "Orthodox axis" embracing Greece, Serbia, and Russia. Turkey also alleges a Greek role in support of the PKK, raising the possibility of Greek-Turkish competition moving into the realm of state-sponsored terror. As a result, long-standing issues are now imbedded in a wider sense of geopolitical rivalry.

Fourth, and very significantly, the military balance in the eastern Mediterranean has been shifting in Turkey's favor—especially in the air.[14] The balance in ground and amphibious forces has long been skewed in Turkey's favor. Only at sea is the balance more nearly even. Both countries have ambitious modernization plans and are acquiring a greater capacity for mobility and longer-range strike. A military clash between Greece and Turkey today would have far greater potential for destructiveness and escalation than in previous decades. As Ankara faces an array of security risks on its borders, apart from Greece, there will be continuing incentives for military modernization, but with inevitable spillover effects on the balance in the Aegean. The Turkish defense-industrial, intelligence, and training relationship with Israel is a new element in this calculus.

It is arguable that since the Turkish intervention of 1974, Cyprus has been more of a political than a security issue in Greek-Turkish relations. Recent developments have combined to make Cyprus once again a central problem in the eastern Mediterranean. The establishment of a joint Cypriot-Greek defense doctrine has had the effect of tying Cyprus firmly into the broader bilateral competition. There is now a very real possibility that the "Turkish Republic of Northern Cyprus" will respond to the prospect of Cypriot EU membership by annexing itself to Turkey, effectively eliminating any possibility of a settlement on the island, and possibly complicating Ankara's own membership bid. Finally, the Cyprus situation has become more heavily militarized. The large-scale Turkish Army presence remains. The Greek

---

[13] Traditional issues in the dispute are surveyed in Monteagle Stearns, *Entangled Allies*, Council on Foreign Relations, New York, 1992; see also Clement H. Dodd, *The Cyprus Imbroglio*, The Eothen Press, Huntingdon, UK, 1998.

[14] For a Greek view of the military balance, see *White Paper of the Hellenic Armed Forces*, 1996–97, Hellenic Ministry of National Defense, Athens, 1997.

Cypriot National Guard has itself acquired more modern equipment, including Russian tanks.

From the Turkish perspective, the Russian role is central. Observers in Ankara do not accept that the transfer of weapons to Cyprus is simply a hard-currency transaction for Moscow. Rather, it is seen as part of a broader Russian strategy of influence and presence in the eastern Mediterranean, with the basic motivation of pressuring Turkey.[15]

Apart from the tangible risk of a Greek-Turkish clash if the Russian S-300 missiles had been deployed as planned, the transfer of this system would also have had implications for U.S. and Israeli security interests in the region. The radar system associated with the S-300s would be capable of monitoring the air space over a large part of the eastern Mediterranean. Information obtained might find its way to Russia, and perhaps Syria or Iran, and could complicate U.S. air operations in the event of an eastern Mediterranean or Middle Eastern conflict.

A Greek-Turkish clash—over Cyprus, or the Aegean, or over ethnic conflict in Thrace—would have profound implications for Turkey and the West. It would also have operational consequences for the United States. In strategic terms, a conflict under current conditions might result in the open-ended estrangement of Turkey from the West, an even more serious situation than that which followed the 1974 events on Cyprus. In 1974, Cold War imperatives argued for restraint in sanctions against Ankara. Today, no such constraints exist, and European opinion, in particular, is likely to be strongly critical of Turkey regardless of the circumstances surrounding a clash. The process of NATO adaptation would also be dealt a blow. New command arrangements in the south would become unworkable. Further enlargement of the Alliance would become difficult against the background of a Greek-Turkish conflict, and failure to prevent, much less contain, a clash would be regarded as a major failure for the Alliance. More broadly, a Greek-Turkish conflict might encourage "civilizational" cleavages in the Balkans and elsewhere. Even Israel might be sensitive to the political consequences of too overt a military relation-

---

[15] Turkish General Staff (TGS) officials point to the presence of some 30,000 to 40,000 Russians on Cyprus as evidence of this trend.

ship in the context of a conflict over Cyprus, especially if Israeli weapons were used, and might look for ways to scale back its cooperation.

The operational consequences could be no less severe. U.S. and allied forces in the eastern Mediterranean might find themselves in harm's way. Criticism of Ankara from Washington or in NATO—a virtual certainty—might cause the Turkish leadership to terminate or suspend key aspects of bilateral defense cooperation and could produce a forced withdrawal from Incirlik. A Greek-Turkish clash would almost certainly provoke active Allied efforts to contain the crisis, negotiate a disengagement, or introduce peacekeeping arrangements. U.S. air and naval forces would likely be called upon to assist in monitoring or separating the combatants.

The risk of a clash and the likely strategic and operational consequences make risk reduction an imperative for NATO, the EU, and the United States. Much of the day-to-day risk in Greek-Turkish relations now stems from air operations, whether in the Aegean or over Cyprus. The air balance is increasingly central to strategic perceptions on all sides. Turkey has made air force modernization a priority, and air power has been the leading vehicle for Turkish assertiveness in the Aegean. So too, in Greece, the air force has emerged as the "hard line" service in the perception of foreign observers, reflecting the reality of daily confrontations in the air and a changing air power balance. As both states acquire large inventories of capable aircraft, as well as new command and control and refueling capabilities, the air dimension can only loom larger in the regional balance. It is therefore worth considering what direct role the United States Air Force Europe (USAFE) and AIRSOUTH might play in risk-reduction efforts, perhaps at the tactical level (e.g., exchanges and pilot-to-pilot meetings).[16]

More positively, and despite U.S. concerns about regional escalation, the Kosovo crisis did not produce new Greek-Turkish tensions. The crisis in fact produced some limited cooperation between Athens and Ankara. Overall, both countries have adopted a relatively cautious

---

[16] Cold War era agreements with the Soviet Union on reducing the potential for incidents in the air could provide useful benchmarks. Confidence-building measures discussed among Arab and Israeli negotiators in the context of ACRS (the multilateral arms control and regional security talks) could provide another.

and multilateral approach to the Balkans.[17] The "earthquake diplomacy" of 1999, coupled with the results of the EU's Helsinki summit, have given rise to a more relaxed atmosphere—without resolving the underlying sources of friction. But strategic dialogue, confidence-building, and risk reduction measures should now find more fertile ground.

TURKISH-SYRIAN CONFLICT: OUTLOOK AND CONSEQUENCES

In recent years there has been a marked increase in Turkish attention to Syria as a security challenge. Indeed, the concern over Syria and the development of a more assertive strategy toward Damascus have been notable changes on the Turkish defense scene, changes that have until very recently drawn little attention in the West. A good expression of Turkish concerns was contained in a well-known analysis by Ambassador Sukru Elekdag, who spoke of a "two-and-a-half war strategy" for Turkey, in which the risk of conflict with Syria figured prominently.[18]

Numerous issues are on the agenda with Syria—disputes over Tigris and Euphrates waters, continued Syrian claims on the Turkish province of Hatay, Syrian criticism of Ankara's relations with Israel, weapons of mass destruction, and above all, Syrian support for the PKK. The last has been a proximate and serious source of risk. Ankara has periodically threatened to strike PKK camps in Syrian-controlled parts of the Beka Valley in Lebanon. There has also been a continuing potential for hot-pursuit incidents between Syrian and Turkish forces pursuing PKK guerrillas on the border. PKK leader Abdullah Ocalan had been based in Damascus, and Syria facilitated PKK operations in Turkey both financially and logistically. In the fall of 1998, Ankara made it clear that Syrian support for the PKK would no longer be tol-

---

[17] Greece and Turkey are playing a leading role in the formation of a Balkan peacekeeping brigade. Greece allowed Turkish aircraft to transit Greek air space in support of humanitarian operations in Macedonia and Albania.

[18] See Ambassador Sukru Elekdag, "Two and a Half War Strategy," *Perceptions* (Ankara), March–May 1996.

erated.[19] With a growing capacity for mobile operations, experience from years of cross-border campaigns in northern Iraq, and against a background of military cooperation with Israel, Turkish decisionmakers appeared confident in threatening military action against Syria.[20] The departure of Ocalan from Syria in December 1998 under strong Turkish pressure—and his apprehension in Kenya in February 1999—is evidence of Turkey's new-found willingness to use its regional weight and operational capabilities abroad. The October 1998 "Adana Agreement" called for the end of Syrian support for the PKK and put in place a monitoring arrangement. Turkish officials are reportedly confident that PKK activity in Syria has been much reduced, although activity in Syrian-controlled areas of Lebanon persists.

A serious Turkish-Syrian clash would have significant consequences. A large-scale intervention aimed at toppling Assad is unlikely, but an unequivocal Syrian defeat on the ground could well weaken Assad's leadership and perhaps change the dynamics in the Middle East peace process (which may have been part of Assad's calculus in agreeing to Ocalan's departure for Moscow and, eventually, elsewhere). Open conflict, or even a protracted period of brinksmanship with Syria, could cause Ankara to seek NATO backing on the basis that terrorists should not be allowed a sanctuary, and that the territory of a NATO member is threatened by Syrian behavior. Given the controversy over the PKK and Kurdish issues in Europe, many allies are likely to balk at the prospect of support for Ankara. NATO's failure to provide a determined response would strongly reinforce existing Turkish concerns about "selective solidarity," first raised during the Gulf War when Germany appeared reluctant to provide ACE (Allied Command Europe) Mobile Force (AMF)-Air reinforcements to Turkey. Sensitivities about Syria's role in the peace process and congressional

---

[19] On October 6, 1998, Prime Minister Yilmaz issued what he described as a "last warning" to Syria over support for the PKK. The TGS and the Turkish parliament have issued similar warnings, and relations have recently been described as a virtual state of war. See "Turks Give Syria Last Warning," *Washington Post*, October 7, 1998; and Howard Schneider, "Turkish Parliament Threatens Syria Anew," *Washington Post*, October 8, 1998.

[20] The military gap between Syria and Turkey is large and growing. Turkey's ground forces are roughly twice the size, and many of Syria's high-grade units are tied down on the border with Israel. In the air, Turkey enjoys considerable superiority with its large inventory of F-16s and other capable aircraft. Syria has perhaps 40 modern, operational fighters (MiG-29 and Su-24). See analysis in Alan Makovsky and Michael Eisenstadt, "Turkish-Syrian Relations: A Crisis Delayed?" Washington Institute, *Policy Watch*, No. 345, October 14, 1998.

concerns might also complicate the response from Washington. Even Israel, generally interested in pressuring Syria, might not find an open conflict in its interest, especially if there is movement in the peace process.[21]

If significant Syrian territory is lost or the survival of the Assad regime is threatened, it is not beyond imagining that Syria might employ Scud B and C missiles against Turkish targets, possibly including Ankara. Adana and Iskenderun would be particularly vulnerable. In this case, the prospects for escalation would increase, as would the incentives for Turkey to explore future deterrent strategies outside a NATO framework. The issue of NATO's exposure to WMD and missile risks would acquire a dramatic and tangible quality.

In sum, Ankara is well placed to achieve an operational success, but conflict with Syria could weaken, rather than strengthen, Turkish ties with the West. In the worst case, perceived abandonment by NATO could produce a crisis in relations with the Alliance.

Confrontation with Syria would be an important test for the Alliance, and will be seen as an Article V rather than an Article VI commitment. Just as promoting confidence-building measures and strategic dialogue between Greece and Turkey is strongly in the Alliance interest, the United States and Europe can play a critical role with Syria by bringing pressure on Damascus to ensure that its disavowal of the PKK is permanent, and that any future Israeli-Syrian disengagement does not increase the risk to Ankara.[22]

THE COMPETITION WITH RUSSIA

Turkey and Russia no longer share a border, but relations between these two historic competitors are, in many respects, less stable

---

[21] Israel was reportedly less than enthusiastic about providing intelligence and other assistance to Turkey in its confrontation with Syria. See "Levantine Labyrinths: A Mini War Between Turkey and Syria Cannot be Excluded," *Foreign Report*, No. 2515, October 6, 1998, p. 1.

[22] Many of the considerations noted in relation to Syria could also apply in the event of a PKK-related clash with Iran. This prospect—once remote—has come to the fore as a result of cross-border incidents and increased tension between Ankara and Tehran in the summer of 1999. Some analysts now suggest that Iran is set to replace Syria as a leading regional sponsor of the PKK, and has become a planning factor for the Turkish military. See Alan Makovsky, "Turkish-Iranian Tension: A New Regional Flashpoint?" *Policywatch*, No. 404, August 10, 1999.

today than they were during the Cold War. In general, the Turkish security establishment's concerns about Russia are focused on longer-term issues: pipeline geopolitics, ethnic conflict and political vacuums in the Caucasus, and the possibility that Turkey might have to face the military risks of a resurgent Russia alone.[23] In particular, Ankara is concerned that NATO, as a whole, will prefer to purchase room for maneuver on further NATO enlargement, Balkan policy, and other controversial issues by allowing Moscow a free hand in dealing with the near-abroad and its southern periphery. Turkish planners fear that Russia will exploit CFE treaty adjustments to rebuild its military potential opposite Turkey.[24] The renewed emphasis on nuclear weapons in Russian military doctrine is another source of concern, encouraging a very conservative attitude toward NATO nuclear policy in Ankara. The sum of these concerns produces a striking degree of wariness about Russian intentions, even as Turkish relations with Russia have expanded dramatically in the economic sphere.[25]

More recently, frictions with Russia in the security realm have acquired a more direct and near-term quality. Russian arms transfers to Cyprus and Iran are the key elements in this regard, but Turks are also concerned about the potential for Russia to play a destabilizing role in the Balkans and the eastern Mediterranean as a whole. Economic and political crises in Russia and the conflict in Chechnya also compel Turks to consider the spillover effects of widespread turmoil across the Black Sea, from refugees to disruption in energy and trade flows. This is an area in which Ankara would like to see greater NATO presence and activity, especially given Partnership for Peace (PfP) activities which, as Turkish naval officials are keen to stress, make the Black Sea a "NATO sea."

---

[23] For a Russian perspective, see Nicolai A. Kovalsky (ed.), *Europe, the Mediterranean, Russia: Perception of Strategies*, Russian Academy of Sciences/Interdialect, Moscow, 1998.

[24] A senior Turkish TGS official described CFE changes as a "green light" to Russia to rebuild its military capability opposite Turkey. General Cevik Bir, "Turkey's Role in the New World Order: New Challenges," *Strategic Forum*, NDU-INSS, No. 135, February 1998.

[25] In the early 1990s, it was fashionable to speculate about vast new economic and political opportunities for Turkey in the Turkic republics of the former Soviet Union. In reality, Russia itself, rather than Central Asia, has emerged as a leading economic partner for Ankara. Russia is now Turkey's leading trade partner (prior to the Gulf War, it had been Iraq), largely the result of energy imports.

THE OUTLOOK FOR RELATIONS WITH THE UNITED STATES AND NATO

The changing state of Turkish-EU relations and new European defense initiatives place greater pressure on the bilateral relationship with Washington, as well as on relations within NATO—the key badge of Turkish membership in the West. Observers on all sides acknowledge that the relationship with the United States has experienced repeated strains in the period since the Gulf War. In the absence of a substantial redefinition of the rationale for the "strategic relationship" between Ankara and Washington, policy differences over Iraq, Iran, the Aegean, and human rights, and democratization issues within Turkey itself, have come to dominate the agenda. The political turmoil in Turkey in recent years has complicated Turkey's ability to engage the West in positive ways. But it has also brought unprecedented attention to relations with Turkey and has given Turkish policymakers a strong stake in repairing the country's image and rebuilding the strategic relationship. Turkey's contributions to Kosovo air operations and KFOR have had a particularly positive effect in this regard.

Turkey is a leading "consumer" of security in the new NATO— many key Alliance planning contingencies involve Turkey in some fashion. TGS strategists themselves note that Turkey's role has changed from a "flank" to a "front." Turkey and Turkish facilities can also play a critical, possibly unique role in Alliance power projection from the Black Sea and the Caspian to the Gulf. But these advantages, conferred by geography, are only theoretical in the absence of a shared strategy toward these regions. As Turkish military and civilian decisionmakers have become more attuned to sovereignty concerns and Turkey's own security interests, the prospects for security cooperation with Turkey have become less predictable, even as a changed strategic environment has increased the utility of Turkish bases from the perspective of Western planners.[26]

In recent confrontations with Iraq, Ankara has been tolerant but unenthusiastic about allowing the use of Incirlik for offensive air op-

---

[26] Discussions at the 1998 USAF Global Engagement "Policy" Game emphasized the importance of Turkish cooperation in Caspian and Gulf contingencies. See United States Air Force, *Global Engagement 98 Policy Game After Action Report*, 9–11 June 1998.

erations in the Gulf. Even the strikes by Allied aircraft within the rules of engagement of Operation Northern Watch are viewed with concern by Turkey's civilian leadership. Overall, the Turkish calculus has been to avoid the unpredictable internal and regional consequences of too-close involvement in the U.S. conflict with Iraq. Ankara is concerned about Iraqi intentions, and is certainly concerned about WMD and missile risks emanating from Iraq. But the political risks of putting Incirlik at the disposal of U.S. forces (i.e., beyond Northern Watch and NATO tasks) will only be tolerable in relation to operations aimed at producing fundamental change in Baghdad or reshaping the security situation in the northern Gulf. In these cases, Ankara would almost certainly want "a seat at the table" and could envision a more-forward-leaning role, as in the Gulf War. Ankara is also sensitive to operations that might embolden rather than contain Kurdish movements in northern Iraq.

The transition from Operation Provide Comfort to Operation Northern Watch (i.e., the end of the ground operation in the north) improved the climate for cooperation and political acceptance of operations at Incirlik. Despite recent labor disputes and some continuing difficulties in day-to-day relations, the operational outlook has improved in the context of Northern Watch. The training environment has become somewhat more permissive, with improved access to the Konya training range for U.S. aircraft. Yet the prospects for the use of Incirlik beyond Northern Watch, and for non-NATO purposes, remain uncertain. The use of Turkish bases in the latter stages of the Kosovo air campaign, a NATO operation, may signal an expanding Turkish view of how their assets and territory may be used for crisis management.

Even though Turkey's own debate over foreign and security policy has become more active and more complex, basic issues such as arms transfers remain key measures of the bilateral relationship in Turkish perception. Some recent arms transfer successes have contributed to a positive climate, and the Turkish military continues to prefer U.S. equipment. But Turkish policymakers still tend to regard congressional scrutiny of arms transfers as a de facto embargo and have taken steps to diversify the country's defense-industrial relationships (Israel, France, and even Russia figure prominently in this regard). Pending purchases of attack helicopters and co-production of main battle tanks will be major tests of the arms transfer climate in the wake of the release of U.S. frigates, a development applauded by Turks but also acknowledged as a fortuitous result of congressional bargaining. The end of any formal U.S. aid to Turkey has also recast the question of what is meant by "best efforts" in the bilateral Defense and Economic Cooperation Agreement (DECA).

A key future challenge will be to involve Turkey in a more predictable fashion in U.S. and NATO strategy toward the European periphery—i.e., security management in the critical regions adjacent to Turkey. To do so, it will be necessary to recast Turkish-western security relations to address the new transregional problems—proliferation, terrorism, refugees, and energy security are prominent examples—facing Turkey and the Alliance. It will also be necessary to accomplish this without reducing Turkish confidence in NATO as a security guarantor in relation to traditional Article V risks, especially from Russia, Syria, and Iran. In the balance of bilateral and NATO approaches to Turkey, it will be useful to emphasize the Alliance dimension wherever possible. Turkish military officials themselves emphasize the importance of NATO activities as a way of engaging ("re-engaging" may be a more accurate term) Europeans in Turkey's interest. This link is likely to acquire greater importance as Ankara seeks to assure itself of participation in emerging EU defense efforts. As Turkey's own sizable military establishment continues to modernize and become more capable of power projection missions, the United States and NATO may look to Ankara to play a direct role in Alliance tasks beyond territorial defense and the provision of well-located facilities.

For efforts at bolstering strategic cooperation to be successful, key near-term risks must be contained. It will be difficult, perhaps impossible, to preserve a legitimate role for Turkey in European security in the event of a conflict over Cyprus or the Aegean. A clash with any of Turkey's Middle Eastern neighbors, in which NATO and EU support is not forthcoming, would similarly jeopardize prospects for engaging Turkey in Western security strategies in the Balkans, Caspian, or the Gulf.

Chapter 5
NATO Adaptation and the South

The security environment around the Mediterranean basin and beyond will be strongly affected by, and will also affect, the process of NATO adaptation. In this context, the adaptation process is understood to include changes in the membership, strategy, missions, and command structure of the Alliance. It also embraces the broader process of change in political influence and roles within NATO. To the extent that the Alliance moves further in the direction of the defense of common interests, with less narrowly defined notions of its security space, Europe's southern periphery should, and will, figure more prominently in NATO strategy. Similarly, as even traditional Alliance missions acquire a greater power-projection flavor, and to the extent that the capacity for power projection becomes more widely shared among allies, the contribution of Southern Region states to NATO objectives is likely to grow.

NATO's Southern Periphery: Alternative Models

As noted earlier, the role of the Mediterranean and adjacent areas in transatlantic security is changing. From Cold War marginalization, the southern periphery has clearly moved to the center of Alliance concerns, even if this significance is rarely defined in "Mediterranean" terms. It is simply that the United States and its European allies are more occupied with southern problems—from the Balkans to the Caspian and from North Africa to the Gulf—and doing more in political and military terms around the region. A number of possible future models for the role of the southern periphery in NATO strategy can be offered, ranging from a simple extension of the current approach to more ambitious concepts that would require a fundamental change in the role of the Alliance.

43

NATO's Near Abroad

A first and least ambitious model treats the southern periphery—broadly, the Mediterranean and perhaps the Black Sea—as essentially an extension of the traditional European security environment; in short, NATO's near abroad. In this model, the emphasis is on a limited expansion of the geographic scope of the Alliance. It takes into account the need for common approaches to new risks, emanating from the south and capable of affecting core European security interests. The key areas of regional concern have been the Balkans and North Africa, and it can be argued that Alliance views have evolved to the point that these places are no longer really "out-of-area." In functional terms, the focus has been on proliferation, terrorism, and refugee movements. This is a balanced model from a transatlantic perspective; crises in this framework are not far from western Europe and may be shaped by European diplomacy and addressed largely (although, as Bosnia and Kosovo demonstrate, not solely) by European military power. The internal problem of Turkish-Greek relations also falls within this frame. Turkey plays an eccentric role in this model of the southern periphery because it is relevant to many potential contingencies and "soft" security problems around the Mediterranean, but many of its own security concerns go well beyond the reach (and in some cases, willingness) of European power. In fundamental respects, this "near-abroad" model of the south is the prevailing model within the Alliance today. Treatment of Mediterranean and southeast European security in NATO's new Strategic Concept falls within this limited rubric.

North-South Security Relations

A second model, and one that has developed in parallel, treats the southern periphery as a theater for north-south relations in security terms. The focus in this approach is on dialogue and forestalling frictions along "civilizational" or "have and have not" lines. Central to this approach is NATO's ongoing Mediterranean Initiative, aimed at dialogue and information-sharing with selected partner states across the Mediterranean, and championed by Southern Region states (especially Portugal, Spain, and Italy). France, although heavily engaged in

Mediterranean and north-south relations, is reluctant to see these relations focused within NATO. The Initiative has evolved from and bears a close relationship to various other Mediterranean initiatives, past and current.[1]

This model is particularly attractive to those within the Alliance, such as Spain, concerned about the consequences of a defense-oriented approach to the south that might be seen across the Mediterranean as a new cold war along north-south, or worse, Muslim-Western lines. As NATO begins to treat defense-related problems in the south more seriously, this "dialogue" model can play a useful confidence-building role. It might eventually become a vehicle for more concrete security cooperation along north-south lines if the Middle East peace process continues to evolve positively.[2] In historical terms, the coexistence of this approach with the first model, described above, is similar to the "Harmel" strategy adopted in relations with the Soviet Union—defense and dialogue in parallel. This is a useful approach to reconciling the looming tension between dialogue and defense in NATO southern strategy. As part of an agreed NATO initiative, the dialogue model enjoys a basic level of political support within the Alliance, although some major and potentially controversial choices will need to be made about how to operationalize the Initiative in the future.

POWER PROJECTION

A third, "power-projection" model views the southern periphery as a logistical anteroom to critical regions beyond the Mediterranean basin—above all, the Gulf and the Caspian. This is a more ambitious conception of the strategic role of the south, going considerably beyond the consensus view within the Alliance. In transatlantic terms, it is heavily weighted toward an American world view and the requirements of U.S. national security strategy. With the limited exceptions of France and Britain, it is also a model that is relevant only in the context of

---

[1] Including the proposed Conference on Security and Cooperation in the Mediterranean (CSCM); the Five plus Five dialogue between the Arab Maghreb Union and southern European states; the Mediterranean Forum; and the EU's Euro-Mediterranean Partnership (Barcelona) program.

[2] See Lesser et al., MR-1164-SMD, 1999.

American power-projection capabilities. It draws heavily on the Gulf War experience in which the Mediterranean served as a critical rear area and in which Southern Region members played an integral part in access and overflight. This is also a model in which proliferation trends in the south and the potential for asymmetric strategies, including terrorism, play a complicating role. The power-projection model highlights the strategic importance of Turkey as a facilitator but also as an increasingly assertive regional actor in its own right. It is also worth noting that some traditional missions for the Alliance, including countering a resurgent Russia in relation to Turkey, are by virtue of their distance part of this power-projection model. In addition to Turkey, this model could argue for more active NATO cooperation with Israel and Jordan as part of a "northern" approach to defense of the Gulf.

Toward a Global NATO?

A final model would treat strategy toward NATO's south as a step toward a more global NATO, with a firm focus on defense of common interests without reference to geographic boundaries. This approach is clearly far beyond the current limits of NATO consensus, but it is not inconceivable over the longer term. It can be regarded as power projection "plus," in the political as well as the operational sense. A more limited conception might see this "global" model as the goal, but with the Mediterranean—Europe's doorstep—as the logical place to start on an expanded transatlantic security agenda.

In the wake of the 1999 Washington summit and the elaboration of a new Strategic Concept, the notion of Europe's southern periphery as NATO's "near abroad," together with the established "dialogue" model, will likely guide Alliance strategy toward the south. But from the perspective of U.S. security interests, and with the likelihood that Alliance members will (perhaps simply as coalitions of the willing) be called upon to act beyond the Mediterranean basin in the future, more ambitious models are also useful. A transforming development, such as the return of France to NATO's integrated military structure, could make more expansive visions of the south's role in NATO strategy viable.

SOUTHERN REGION PERSPECTIVES ON A CHANGING NATO

Perspectives on NATO adaptation vary considerably from Lisbon to Ankara, and it would be misleading to attempt a synthesis. This analysis is aimed at characterizing the key lines of interest and concern, particularly in relation to future NATO missions. Although not a traditional "Southern Region" state, France's views are reflected here as a key actor in Mediterranean security and a strong influence over strategic thinking in southern Europe.

Future of NATO's Mediterranean Initiative

There is an unresolved tension in Southern Region attitudes toward the south. Policymakers and observers applaud the new attention to Mediterranean security problems, and some states, particularly in the western Mediterranean, believe that dialogue and confidence-building, rather than new defense initiatives, should be the centerpiece of NATO strategy toward the region. By contrast, Turkey is more comfortable with a defense-oriented approach. NATO's Mediterranean Initiative is seen as worthwhile, but it has suffered from a lack of resources and an inability to bring the dialogue states, including Israel, together in a true multilateral fashion.[3] At the moment, the Initiative is almost exclusively bilateral in character, so opportunities for risk-reduction in a south-south context are lost.[4] There is little support for the idea of a formal "partnership for peace" program in the Mediterranean (an idea first raised by the former Italian Chief of Defense Staff). But the Initiative might be given a boost through new PfP-like activities around the Mediterranean, perhaps giving some existing bilateral exercises a NATO "hat." With its active program of bilateral military activities in the Mediterranean (e.g., with Egypt), the United States could play a key role in this regard. There is a consensus that renewed progress in the Middle East peace process—a more tangible prospect in the wake of

---

[3] As an example, places have been reserved for dialogue partners at the NATO Defense College in Rome and at the NATO school in Oberamergau, but these places are self-funding. The same is true of invitations to observe NATO exercises.

[4] See Claire Spencer, "Building Confidence in the Mediterranean," *Mediterranean Politics*, Vol. 2, No. 2, Autumn 1997; see also a forthcoming report for the U.S. Institute of Peace by Roberto Aliboni.

the 1999 Israeli elections and the revival of Syrian-Israeli negotiations—would transform the climate for the Initiative.

Functional Versus Geographic Missions in the South

French and Spanish analysts see greater attention to Mediterranean challenges as a key step toward ensuring the continued relevance of the Alliance in a changing security environment. But these states do not wish to see the Mediterranean region "singularized" as an area of threat requiring special treatment. The concern is two-fold. Neither country wishes to complicate delicate political relationships across the Mediterranean through a more assertive declaratory strategy toward the south.[5] For France, under current conditions at least, there is also little interest in seeing NATO become the centerpiece for Western strategy toward the Mediterranean.

The new Strategic Concept notes, but does not stress, the role of the "Mediterranean" in Alliance strategy.[6] However, the region's future importance to the Alliance is strongly defined in functional rather than geographic terms, that is, in terms of new missions (peacekeeping, crisis management, counterproliferation, etc.). These missions will be inherently "southern" in character and far more likely to be carried out around the Mediterranean than on the Polish border. Power-projection missions are similarly seen as most likely and most demanding in relation to crises on the southern periphery.

Southern Region observers as well as AFSOUTH officials point to the looming gap between planning and operational demands in the south, including ongoing requirements in the Balkans, and the NATO resources traditionally devoted to Southern Region military activities and infrastructure. Southern Region contingencies dominate post–Cold War NATO military planning, but the region has perhaps 20–25 percent of total NATO assets and activity.[7] Southern Region infrastructure, whether funded nationally or through NATO infrastructure funds, is widely seen as undercapitalized. As noted earlier, this under-

---

[5] The delicate issue for France is Algeria; for Spain it is the enclaves of Ceuta and Melilla.

[6] See NATO document NAC-S(99)65, 24 April 1999, paragraph 38.

[7] This is an AFSOUTH estimate. Admittedly, it is a difficult measurement to make, but the rough percentage is illustrative of a perceived imbalance.

capitalization weighs heavily on NATO's ability to address proliferation risks in the south.

Traditional (Article V) Versus Nontraditional Missions

There is a notable divide within the Southern Region on the question of traditional Article V missions oriented toward the defense of territory versus nontraditional missions aimed at the defense of common
interests. In reality, of course, there is no fundamental conflict between these missions because there is no suggestion that the Alliance
abandon Article V commitments. However, Southern Region states are
especially sensitive to the longer-term implications of shifts in emphasis, perhaps because their security concerns have long been at the mar-
gins of NATO strategy.

Portugal, Spain, and Italy have been at the forefront in attempting
to reorient Alliance strategy toward security risks beyond territorial defense. There is also a strong Spanish and Italian interest in engaging the
Alliance on what might otherwise be viewed as "nonshared" risks.[8]
Madrid would certainly favor any evolution of the Alliance concept
that strengthened the outlook for a multilateral approach to the security of the enclaves of Ceuta and Melilla on the Moroccan coast. Italy
has a similarly strong interest in NATO support in dealing with refugee
flows from across the Adriatic or from North Africa. Southern Europe
generally is interested in additional reassurance on energy security issues.

Without dismissing the significance of nontraditional challenges
and missions, Greece and Turkey share a more conservative view of Alliance missions. Both Athens and Ankara continue to regard threats to
borders as a serious concern. The Kosovo experience reinforces this
view. Turkey faces a host of potential threats to its territorial integrity,
as well as proximate risks from ballistic missiles and insurgents—both
definable as Article V-type problems. Both are especially concerned
about reaffirming Article V commitments in light of the enlargement
process, which is widely seen—rightly or wrongly—as introducing a
new spirit of conditionality in Alliance security commitments.

---

[8] The term is Spanish, and is usually applied in relation to the defense of Ceuta and Melilla—within the
WEU area but outside of NATO.

There are also some notable differences in perspective on possible new missions envisioned in the Strategic Concept—in particular, crime, drugs, and terrorism. These so-called "third pillar" issues (to use EU terminology), traditionally regarded as national responsibilities within the Alliance, inspire varying responses across the Southern Region. In the western Mediterranean, inclusion of these issues is generally non-controversial, except in France where there is a strong preference for EU-based approaches. Greece too is reluctant to see NATO undertake initiatives on terrorism or international crime, apart from drug trafficking.[9] Given the struggle against the PKK, Turkey has a strong interest in seeing counterterrorism emerge as a new NATO mission, but is wary of initiatives on crime and drug trafficking, where the Turkish experience is controversial.

Outlook and Preferences on Enlargement

There has been an evolution in Southern Region views on NATO enlargement.[10] Early in the enlargement debate, southern European and Turkish perceptions could fairly be characterized as neutral at best, and often negative. Concerns centered on the likely dilution of attention and resources, and the migration of Alliance influence eastward—concerns shared elsewhere, but with particular relevance for smaller allies in the south. Turkey, with multiple—and controversial—security problems on its borders, also feared that a larger NATO would be a NATO with more conditional, less automatic security guarantees. Overall, Southern Region opinion is now more positively disposed toward the enlargement process, although there is still some sensitivity in the eastern Mediterranean about the longer-term effect on security guarantees. Greece and Turkey, in particular, have come to accept arguments about the stabilizing contribution of NATO membership in regional security. Analysts in both countries also urge that NATO extend this argument to the insecure Balkans, where their own interests are di-

---

[9] Persistent left-wing terrorism and Greek policy toward international terrorist activity in Greece has been a consistent issue in bilateral relations between Washington and Athens. Greece and Cyprus have also been criticized for lax policy toward money laundering. These are now signs of improvement in all of these areas.

[10] Maurizio Cremasco provides an Italian view in "NATO Enlargement in Light of the Madrid Decision," unpublished draft, February 1998.

rectly engaged (both Athens and Ankara are also keen to play a more active role in PfP activities in the region).

Across the region there is an expectation of, and support for, the idea that subsequent enlargement moves include southeastern Europe. The most promising candidates in this regard are Slovenia and Romania. A southward enlargement would help "secure the Balkan hinterland" and would encourage a geographically (read *politically*) balanced NATO. This last consideration reflects the widespread concern in NATO's south that the accession of Poland, Hungary, and the Czech Republic strengthens German influence in European and transatlantic affairs, and reinforces an eastward-looking bias in NATO strategy.[11] Romania might also be a useful partner for an Alliance that becomes more interested in power projection toward the Black Sea and the Caspian.[12]

Nuclear Policy

Overall Southern Region opinion is relatively relaxed about nuclear weapons and strategy, although, with the notable exception of Turkey, there is continuing interest in the political benefits of a less nuclear NATO. This argument extends, especially in Spain, to the benefits for bilateral security cooperation with the United States of further reductions in nuclear weapons based in Europe. The new Italian government, with its leftist background, faces different dilemmas on nuclear issues. Rome is keen to demonstrate its reliability as a NATO ally, but has some sympathy for antinuclear sentiments emanating from Germany. Privately, some southern European strategists are interested in retaining a robust declaratory strategy and appropriate nuclear systems based in Europe to deter looming proliferation risks. This interest may well become more overt over the next decade. Turkish views are more straightforward, with multiple proliferation risks on Turkey's borders and the persistent problem of Russian nuclear forces (and doctrine) arguing for a strong NATO commitment in this area. Ankara

---

[11] However, NATO will acquire an additional Southern Region member in the process; Hungary will be assigned to ASFSOUTH. If Austria were to become a member, it too would likely become part of NATO's south for command purposes.

[12] Romanian defense officials stress their country's role in facilitating Western power projection to these regions, perhaps in cooperation with Turkey and Israel. See Robert D. Kaplan, "The Fulcrum of Europe," *The Atlantic Monthly*, September 1998.

is also exposed to the consequences of loose nuclear weapons, material, and expertise resulting from chaos in Russia.[13]

Command Reform

Command reforms already in place (including the shift to two regional commands) should bolster the weight of the Southern Region in NATO planning and focus additional attention on risks emanating from the south. The activation of new Joint Sub-Regional Commands (JSRCs) also implies a more active role for Southern Region members in NATO command arrangements. New commands in Verona and Madrid are noncontroversial. New JSRCs at Izmir and Larissa are more controversial, and could easily become embroiled in broader Greek-Turkish tensions, although good progress has been made on this front in Athens and Ankara, and new JSRC arrangements are going forward in the eastern Mediterranean. Outside the AFSOUTH area, but nevertheless part of the security equation along the southern periphery, the restructuring of command responsibilities in the eastern Atlantic to accommodate a more active Spanish role in the Alliance and the prospective establishment of SOUTHLANT will improve NATO's ability to act in West African and North African contingencies. SOUTHLANT along with AFSOUTH can also play a role in exercises, exchanges, and information activities associated with the Mediterranean Initiative.[14]

Turkey, Italy, and Greece have volunteered to serve as "framework nations," providing headquarters and command and control infrastructure for an additional NATO rapid-reaction corps in the south. Under current plans, Hungary will be the only Southern Region country without a NATO command. This could provide a future opportunity to establish a JSRC for the Balkans or the Black Sea, and could facilitate an operational air presence in southeastern Europe.

---

[13] Turkey has been the scene of some prominent attempts to sell or ship nuclear material from the former Soviet Union.

[14] RADM John Paddock, "Cooperation in the Eastern Atlantic and the Role of Iberlant," briefing, September 25, 1998. See also statement by Dr. Jaime Gama, Minister of Foreign Affairs of Portugal, "The Azores and the New Transatlantic Partnership," *Furnas*, September 28, 1998.

Transatlantic Roles, Capabilities, and Mandates

Among the Southern Region states, Portugal, Italy, and Turkey have historically been most concerned about maintaining an active U.S. presence in European security affairs. This pattern can be expected to continue or perhaps be strongly reinforced in the case of Turkey. Turks are wary of any change in NATO that points to more European influence at the expense of U.S. engagement, or promotes the role of European institutions from which Ankara is excluded, or in which its influence is restricted.[15]

Elsewhere, movement toward a common European foreign and security policy, a stronger European defense identity, and generally a more balanced approach to transatlantic security roles is favored. The Combined Joint Task Force (CJTF) concept, in particular, is applauded as a means of giving Europe a greater capacity for crisis management, especially on the southern periphery where many potential European-led, U.S.-supported operations can be envisioned. The critical component in many cases will be U.S. tactical air power as well as airlift. The maintenance of a standing U.S. air presence in or readily accessible to the Southern Region is thus intimately connected with southern European perceptions of the utility of the CJTF model. The confluence of likely movement toward a more power-projection-oriented alliance and European exposure to proliferation risks in the south will probably reinforce this linkage among presence, capability, and reassurance against retaliation.

The Defense Capabilities Initiative outlined at the Washington summit, together with European defense initiatives presented at the EU's Cologne summit and more fully articulated in Helsinki, suggest that Europe may be "getting serious" about acquiring more mobile and capable military forces. Kosovo has given further impetus to this trend. If so, the impact of these new European, including southern European, capabilities will be felt, first, in the potential for intervention in adjacent Mediterranean areas.

Southern Region observers generally favor and anticipate the eventual return of France to the integrated NATO command, and would

---

15 Turkey adopted a tough negotiating position at the Washington summit, withholding agreement on a new Strategic Concept in order to secure guarantees regarding its role in WEU decisionmaking and European Security and Defense Identity (ESDI).

view this development as a critical contribution to the future effectiveness of the Alliance in the south. For some, the attractiveness of this prospect is enhanced by the belief that it would balance a more active German role in peacekeeping and crisis management in the Balkans and the Mediterranean.

One feature of the post–Cold War European security environment troubling to NATO's southern allies—France excluded—has been the rise of contact group formulas in addressing regional crises. There is a widespread belief that this approach tends to marginalize smaller allies, even in those cases where their security interests are directly concerned. Italy's experience of exclusion from Balkan diplomacy in pre-Dayton Accord Bosnia, despite its proximity and critical operational contribution, provides a clear example. A degree of conservatism about out-of-area operations encourages most southern allies to favor reliance on a clear-cut UN mandate, wherever possible, for operations outside the treaty area.

In sum, the process of NATO adaptation promises increased attention to the Mediterranean and its hinterlands, above all, as part of new functional missions for the Alliance that are most likely to be performed on the European periphery. Adaptation will also encourage increased activism and assertiveness on the part of southern allies.

Chapter 6
Conclusions and Policy Implications

Overall Observations

The Washington summit and the Kosovo experience have brought new attention to risks emanating from NATO's southern periphery. The new Strategic Concept identified the Mediterranean as an area of security concern, and the Alliance has reaffirmed its commitment to the existing Mediterranean Initiative. That said, the thrust of Alliance strategy toward the south will be defined in functional rather than geographic terms, with an emphasis on new missions—from countering proliferation risks to crisis management—that are more likely to be performed in the south than elsewhere.

As the Alliance moves to concentrate more on the defense of common interests and power-projection missions, it will naturally focus additional attention on the south—the Southern Region members, the Mediterranean states involved in partnership and dialogue with NATO, and the wider region where developments can affect transatlantic security interests. An evolution in this direction will also serve U.S. strategic interests, encouraging greater European involvement in defense on the periphery, bolstering the relevance of U.S. military presence and strategy in Europe to new transregional security challenges, and contributing to U.S. freedom of action in extra-European crises.

Key security relationships around the Mediterranean, both bilateral and through NATO, have not adjusted to reflect post–Cold War realities. These relationships require redefinition to provide a predictable basis for cooperation in addressing post–Cold War problems. NATO's new Strategic Concept will be helpful in defining a new agenda for defense cooperation, but it is not sufficient in its own right. The United States needs to explore ways of jointly redefining key bilateral rela-

55

tionships in the Southern Region through more frequent high-level interaction with leaderships.

Nowhere is this need for redefinition more acute than in relations with Turkey. Internal uncertainties and multiple security risks (the term "threats" still has relevance for Ankara) make Turkey the new front-line state within the Alliance. But there is no transatlantic consensus on policy toward Turkey. Turkey has emerged as an important but also much more assertive security partner. In the absence of a concerted effort to reengage Ankara in European security affairs and to reassure Turkey about the solidity of the NATO security commitment, the United States and the Alliance risk losing a key asset in shaping the new strategic environment. A new agenda for security relations with Ankara will need to focus on proliferation risks, counterterrorism, and energy security—common interests across the Southern Region. It will also need to address Turkey's special concerns about pressure from a resurgent Russia on the southern periphery.

Failure to address the continuing risk of a Greek-Turkish conflict jeopardizes Alliance adaptation and European security. Implementation of risk-reduction measures, along the lines of agreements brokered by NATO's Secretary General, is imperative. Strategic dialogue to manage longer-term risks, including disputes in the Aegean and Cyprus, should have a broad agenda and could embrace arms control, Balkan and Black Sea reconstruction, and regional crisis management. As a hedge, however, it is essential that the Alliance—or at least key members—develop plans in advance to monitor and contain a possible clash in the eastern Mediterranean. These efforts will be facilitated by recent positive changes in the climate of relations between Athens and Ankara.

Expanded NATO involvement in the Mediterranean—Europe's "near abroad"—is a logical step toward a broader transatlantic security partnership, embracing more ambitious models of strategy toward common security interests in the Gulf, the Caspian, and elsewhere. Germany is emerging as a significant actor in the Mediterranean region and can be a part of this evolution. The return of France as a full NATO partner would be a transforming development in strategy toward the south and should be a priority objective of U.S. policy. The Kosovo experience reinforces these points, and recent decisions regarding a Common Foreign and Security Policy (CFSP) and ESDI could encourage a balanced approach to transatlantic roles on the European periphery. It may also facilitate the eventual reintegration of France in NATO military structures.

Greater attention to the south in Alliance strategy should imply a shift of NATO resources southward. Most, and the most likely, NATO contingencies are in the south, but the vast bulk of Alliance resources remain north of the Alps. Costs associated with the integration of new members in the east will impose competing demands, and a more expeditionary strategy may offset requirements for permanently based assets in the south (there may even be benefits to keeping a relatively large proportion of forces available for use on the periphery in the rear). At a minimum, however, addressing new risks in the south, especially counterproliferation and air defense, will require improvements to the undercapitalized and outdated infrastructure across the Southern Region.

Toward a Southern Strategy for NATO

NATO has taken steps to integrate Mediterranean security concerns and initiatives in its broader strategy. Given the security demands emanating from the region, a more focused strategy toward the south is called for. Such a strategy can be outlined in three dimensions: core objectives, shaping the security environment, and hedging against regional uncertainty.[1]

Core Objectives

The Alliance continues to have important Article V responsibilities in the south, particularly on Turkey's borders. Deterring and defending against these risks to Alliance territory are core objectives of NATO strategy. A second and increasingly prominent core objective will be to defend common interests on Europe's periphery.

---

[1] This tripartite framework for strategic planning has been applied in numerous RAND analyses. See, for example, James Dewar et al., *Assumption-Based Planning: A Planning Tool for Very Uncertain Times*, RAND, MR-114-A, 1993.

Environment Shaping

To help promote NATO's core objectives, NATO strategy needs to address security problems around the Mediterranean in a proactive manner. Key tasks in this regard include the prevention and management of regional crises, including flashpoints in Greek-Turkish relations. Similarly, the Alliance needs to contain new security risks, especially those of a transregional character such as WMD and missile proliferation, spillovers of terrorism and political violence, and threats to energy security. NATO's Mediterranean Initiative can play a vital role in environment shaping by promoting security dialogue and engaging nonmember states in North Africa and the Middle East in defense cooperation, training, and crisis management activities.

Hedging Against Uncertainty

The Mediterranean is a crisis-prone region experiencing rapid change. NATO strategy must anticipate the need to mitigate the effects of regional instability, including consequences that may be felt on NATO territory. Dealing with disastrous refugee flows and civil emergencies will be part of this hedging dimension, as will anticipating and preparing for humanitarian interventions.

IMPLICATIONS FOR MILITARY PLANNING

Power Projection and Demands in the South

Distance, diversity of risks, and Alliance geography give aerospace power a special role on Europe's southern periphery. The AFSOUTH area of regard now stretches from Mauritania and the Canaries to the Caucasus. The extent of this security space and the need for NATO to move toward a greater power-projection orientation suggest that the role of air power in European security has changed significantly in the wake of the Cold War. The key risks in the new European security environment are transregional in nature, which means that the defense of NATO interests may well take place outside the NATO area, and perhaps beyond European territory. European-based air power will likely be called on to a greater extent for interventions outside Europe, in the Middle East and Eurasia. NATO in the new strategic environment is likely to place more, not less, emphasis on air power, and the bulk of future demands across a range of missions— humanitarian assistance, counter-WMD, halting conventional aggression, counterterrorism, crisis management—will emanate from the south.

Supporting Expeditionary Operations in the South: Spain, Italy, and Turkey

Our analysis does not suggest the need for significant re-basing of USAFE air assets.[2] Rather, an expeditionary approach to power projection in NATO's south suggests the importance of reinforcing access arrangements around the Mediterranean. Italy, and particularly Turkey, will be key centers for the projection of air power in the new environment. Italy's proximity to the Balkans and North Africa, and a generally favorable political acceptance climate, give it a special role in facilitating the projection of tactical air power, as well as in supporting airlift and strategic air operations further afield.[3] Preserving this relationship over the longer term may, however, require USAFE to consider ways of transferring some operations from the congested north of Italy to the south. Italy would also be an ideal center for new multinational air operations, possibly in a CJTF context.

Turkey will be critical. Its importance in the power-projection equation will only be enhanced by future concerns about the Caspian, counterproliferation demands, and possible disruptions in traditional approaches to defense in the Gulf. Moreover, some key contingencies for the Alliance will involve the defense of Turkey itself. Turkish constitutional prohibitions prevent the permanent stationing of Allied air forces, but there will be a need for a framework to allow the long-term rotational presence of tactical air power at Incirlik beyond Operation

---

[2] See David Ochmanek, *NATO's Future: Implications for U.S. Military Capabilities and Posture*, RAND, MR-1162-AF, 2000.

[3] Similar conclusions were reached in the context of a 1991 RAND study. See Ian Lesser and Kevin Lewis, *Airpower and Security in NATO's Southern Region: Alternative Concepts for a USAF Facility at Crotone*, RAND, N-3264-AF, 1991. The increase in U.S. air traffic through Sigonella provides additional testimony to the importance of Italy as a logistical line to the Gulf.

Northern Watch. Above all, the USAF access and overflight relationship must be more predictable. Improved military-to-military cooperation can play a role. But translating Turkey's geostrategic advantages into operational benefits can only be accomplished through high-level bilateral agreement on regional defense strategies. Looming U.S. arms transfer decisions concerning attack helicopters, tanks, and other systems will provide a critical context. Failure to conclude these transfers may make more ambitious approaches to strategy and defense cooperation difficult or impossible.

Along with Turkey and Italy, Spain provides another key element in the Southern Region air power equation, especially with regard to airlift. Facilities at Rota are increasingly devoted to air operations, and improvements at Moron will support more use of this facility in future contingencies. Spain has been, and will continue to be, especially important in supporting humanitarian and peacekeeping operations in West Africa and the Maghreb. As in Italy, bilateral cooperation on air issues is generally good, but growing demands for transparency in the use of bases—a product of the general increase in sophistication of the Spanish security debate—can impose constraints. In this setting, the interest of Spanish military and civilian officials in expanded high-level, bilateral strategic planning dialogue becomes significant. Politicians and defense officials will need to portray the need for cooperation in terms that demonstrate not just consultation, but joint strategic interest.

Preserving Military-to-Military Ties

The United States has especially close historical ties to southern European and Turkish air forces. The preference for American equipment, training (senior Southern Region officers tend to have spent time in the United States), and security assistance have provided the bases for close bilateral relations. The end of U.S. security assistance to NATO Southern Region countries is evidence of more mature relationships, but it also removes a key basis for cooperation. New defense-industrial initiatives can play this role where arms transfers are noncontroversial. With key countries such as Greece and Turkey, this is not often the case.

Generational change in Southern Region air forces also raises questions about the future quality of military-to-military cooperation and the outlook of future officers. An increasingly European orientation among most southern allies has encouraged closer defense-industrial and training ties with European partners. In Turkey, the impetus for diversification comes from concerns about the unpredictability of U.S. arms transfers. New engagement efforts through USAFE can help to offset these changes in attitude about bilateral cooperation over the longer term. In broad terms, however, the trend toward diversification has become a part of the environment, and will be reinforced by new EU defense arrangements.

Advantages of a Portfolio Approach to Presence and Access

More expeditionary approaches to power projection and crisis management place a premium on flexibility in access. As the political scene in NATO's south (and among countries outside the Alliance but within the NATO orbit) changes, there will be new opportunities for establishing presence and access relationships. Beyond providing additional operational flexibility and extending air power's reach to new areas of concern such as the Black Sea, a portfolio approach incorporating multiple basing options can increase the predictability of cooperation by reducing the perception that an ally is being "singularized" (a concern in Italy and Turkey). Candidates for augmenting the portfolio include Hungary and Romania for Balkan and Black Sea scenarios. Azerbaijan may be a viable alternative for Caspian deployments. Changing attitudes in Greece might even make Crete attractive for North African contingencies. Existing British bases on Cyprus might also be useful in relation to the Levant and the Gulf. A portfolio approach to access is a useful hedge against uncertainties about coalition behavior in future crises, not least the potential unavailability of transit through the Suez Canal and the consequent increase in airlift requirements.

Anticipating Future Military Contributions

All of NATO's southern allies are in the process of restructuring and modernizing their militaries to create more capable and readily deployable forces. Progress on ESDI and NATO's Defense Capabilities Initiative should give further impetus to this trend, although (with the exception of Turkey and Greece) low levels of defense spending will place limits on future capabilities. The scale of Turkish modernization plans suggests that Turkey will emerge as a very capable regional military power over the next decade. The southern allies are already capable of making significant contributions to amphibious operations in their own subregions (e.g., in North Africa and around the Adriatic). At the same time, and as the Bosnia and Kosovo operations show, the political will exists to use these forces in regional contingencies. Apart from the recognition of new defense requirements, such as maritime surveillance, the ability to contribute to coalition peacekeeping and humanitarian operations is a strong motivating factor behind southern European modernization programs. The growth of a stronger European defense component within NATO will be helpful in this regard, because the European link is increasingly important in justifying costly defense programs. Apart from France, few of NATO's southern allies are likely to be capable of more than symbolic deployments in more distant Gulf or Caspian contingencies. However, many Mediterranean contingencies are "close-in" cases where southern European forces can play a leading role—Operation Alba in Albania provides a good example. Thus, the future importance of southern allies goes beyond basing, overflight, and host-country support.

Increasing the NATO Content of Air Power Activities

Where appropriate, existing bilateral air power activities in the south should be given a NATO flavor. In key Southern Region countries, NATO content can help to reduce political acceptance problems and may help accustom southern allies to more expansive Alliance missions. As controversial as some aspects of the NATO operations in Kosovo were (e.g., in Greece, and to an extent, in Italy), the extensive basing, overflight, and force contributions by southern allies would have been difficult, perhaps impossible, to secure on a bilateral or limited coalition basis.

Outside NATO, and especially with NATO's Mediterranean Initiative partners, some bilateral exercises and other activities might also be conducted "in the spirit of" the Mediterranean Initiative. Morocco, Egypt, Israel, and Jordan would be key candidates in this context.

Imperative of Greek-Turkish Risk Reduction

Greece and Turkey possess highly capable air forces and this capability is set to grow. At the same time, the confrontation over the Aegean and Cyprus is increasingly found in the air. Initiatives aimed at risk reduction and confidence-building in the air can therefore make a disproportionate contribution to stability in the eastern Mediterranean—and will be more practical in the wake of improved political relations between Athens and Ankara. By contrast, open conflict, or even continued brinksmanship, can impose opportunity costs on Alliance effectiveness and U.S. freedom of action in the most risk-prone part of the European environment. The political consequences of a Greek-Turkish clash could, for example, include the open-ended denial of access to the Incirlik air base. Given the nature of the stakes, the United States and NATO should be prepared to contribute air power assets to demilitarization and confidence-building activities, including the monitoring of no-fly zones that might be agreed to as part of future Cyprus or Aegean settlements.

AREAS FOR FUTURE RESEARCH AND ANALYSIS

These findings point to several areas where additional analysis will be useful to U.S. policymakers and USAF planners. Recent developments in the Balkans, the Aegean, and the Middle East, in particular, may offer opportunities to extend U.S. and NATO strategy and to enhance the USAF's capacity for power projection on the European periphery.

Lessons of Kosovo for Basing and Access

The Kosovo air operations depended critically on access to facilities in Italy and benefited from the use of a range of facilities elsewhere in NATO's south. The intensity of air operations from Italy put the issues of political acceptance and congestion associated with Aviano and other facilities in northern Italy in sharper relief. Questions raised by Kosovo include: Should the USAF consider rebasing options elsewhere in Italy? How might such decisions be affected by political-military issues in Italy, NATO programs, budgetary considerations, and operational needs? What approach holds the greatest promise of sustainable access across a range of scenarios? Beyond Italy, has the Kosovo experience opened the door to a wider range of regional basing options (Turkish Thrace, Albania, and other areas)? What are the political-military, resource, and operational implications of a broader reassessment of basing and access options?

Beyond Northern Watch, What Role for Air Power Based in Turkey?

The USAF use of Incirlik and key aspects of the defense relationship with Turkey have been driven by the requirements of Operation Northern Watch. Looking ahead, what should the United States and the USAF seek in terms of access to and use of Turkish facilities? What purposes and arrangements are worth considering, and what can USAF planners anticipate from the Turkish side with regard to conditions of use, labor and host-country support, training, and other matters that have complicated recent relations at Incirlik? Does the Kosovo experience suggest the opportunity and need for access to a wider range of facilities on Turkish territory? How might this be accomplished from a legal and political standpoint? How should a changing force protection climate in Turkey affect the USAF calculus?

Building on the findings of this report, it would be useful to consider in detail the role of Turkish facilities in a "northern approach" to power projection for the Gulf, possibly in conjunction with "post-peace" arrangements involving Israel, Jordan, or others.

Potential USAF Contributions to Greek-Turkish Risk Reduction

This report points to the importance of risk-reduction measures between Greece and Turkey, especially in the air. There is interest in exploring such initiatives on a military-to-military basis on all sides. What concrete measures can be envisioned? What role could USAFE play as a stakeholder and facilitator? In conjunction with NATO and individual allies, what kinds of contingency planning for crisis management in the Aegean are politically and operationally practical? How could USAFE assets contribute?